IMMUNODIAGNOSTICS
PRINCIPLES AND PRACTICE

IMMUNODIAGNOSTICS
PRINCIPLES AND PRACTICE

S.C.Rastogi
Formerly
Professor of Biology
B.I.T.S., Pilani

PUBLISHING FOR ONE WORLD
NEW AGE INTERNATIONAL (P) LIMITED, PUBLISHERS
New Delhi • Bangalore • Chennai • Cochin • Guwahati • Hyderabad
Jalandhar • Kolkata • Lucknow • Mumbai • Ranchi
Visit us at **www.newagepublishers.com**

Published by New Age International (P) Ltd., Publishers
First Edition: 1996
Reprint: 2009

Branches:

- 36, Malikarjuna Temple Street, Opp. ICWA, Basavanagudi, **Bangalore**. ✆ (080) 26677815
- 26, Damodaran Street, T. Nagar, **Chennai**. ✆ (044) 24353401
- CE-39/1016, Carrier Station Road, Ernakulam South, **Cochin.** ✆ (0484) 4051303
- Hemsen Complex, Mohd. Shah Road, Paltan Bazar, Near Starline Hotel, **Guwahati**. ✆ (0361) 2543669
- No. 105, 1st Floor, Madhiray Kaveri Tower, 3-2-19, Azam Jahi Road, Nimboliadda, **Hyderabad**. ✆ (040) 24652456
- RDB Chambers (Formerly Lotus Cinema) 106A,1st Floor, S.N. Banerjee Road, **Kolkata**. ✆ (033) 22275247
- 16-A, Jopling Road, **Lucknow**. ✆ (0522) 2209578
- 142C, Victor House, Ground Floor, N.M. Joshi Marg, Lower Parel, **Mumbai**. ✆ (022) 24927869
- 22, Golden House, Daryaganj, **New Delhi**. ✆ (011) 23262370, 23262368

ISBN: 978-81-224-0908-6

Rs. 395.00

C-09-08-3967

Printed in India at Glorious Printers, Delhi.
Typeset at Jayanti Graphics, Delhi.

PUBLISHING FOR ONE WORLD
NEW AGE INTERNATIONAL (P) LIMITED, PUBLISHERS
4835/24, Ansari Road, Daryaganj, New Delhi-110002
Visit us at **www.newagepublishers.com**

To

The Memory

Of My Parents

Preface

As the world population grows, the need for health care increases. Health care today not only means advances in therapy but also requires accurate, sensitive, specific, quick and cost-effective diagnostic systems. With the advances in biotechnology, it has now become possible to develop diagnostic kits for several diseases and pathological conditions. Clinical biochemistry today is very different from the one which existed a few decades ago. Immunodiagnostics has replaced, and is replacing conventional techniques of serum biochemistry and analysis. With the advent of monoclonal antibodies, immobilised enzymes and membrane biosensors, the scenario of the clinical laboratory has undergone a seachange.

Appearance of certain unusual proteins or antigens in certain parts of the body or in circulation gives rise to several pathological symptoms. In some cases there may be increased or decreased levels of certain substances, such as enzymes, hormones, etc., as compared to the normal levels in the human system. If these substances serving as antigens are quantified, they may yield valuable information about the pathophysiological state of the body. Bioassays utilising antibodies as tools to perform such qualitative, quantitative or semi-quantitative determinations of either clinical or academic interest can be broadly grouped under the class of immunoassays and the subject of their study is referred to as *immunodiagnostics.*

Immunodiagnostic techniques are not only important in medical diagnosis, but also have several applications of academic interest in biological sciences. Immunodiagnostics is an interdisciplinary area requiring inputs from the fields of immunology, molecular biology, biotechnology, physical chemistry, microbiology, biochemistry, electronics and instrumentation.

The purpose of this book is to create an understanding of the principles and the applications of some of the well-known and proven methods of immunodiagnostics. Chapter 1 introduces the subject of immunology and the latest terminologies — one should be acquainted with. Though immunodiagnostics is an interdisciplinary science, its basis is immunology, and a thorough understanding of the molecules that participate in the immune system is obligatory before one proceeds to study its applications.

The underlying principles of immunodiagnosis are essentially those emanating from immunocytochemistry. Immunochemical principles have been discussed in Chapter 2, where a detailed account of the antigen-antibody reaction at the molecular level, its kinetics and implications, are described.

Chapter 3 deals with the techniques in immunodiagnostics known as unlabelled immunoassays. These assays were, in fact, the first immunodiagnostic methods to be developed. Precipitation, agglutination and complement fixation are the techniques that have been discussed, and practical examples of the kits in use have been cited.

Chapter 4 surveys most of the sophisticated applied immunodiagnostic techniques — the labelled assays. Several types of labelled assays, both isotopic and non-isotopic, have been discussed, and specific examples of available kits in medical disgnosis provided.

Immunodiagnosis does not restrict itself to immunochemical reactions in a test tube alone, but also offers scope for detection and quantification of cellular and tissue antigens. This forms the basis of immunocytochemistry, the topic for discussion in Chapter 5. A remarkable feature of this Chapter is the technique of *immunogold labelling*, a recent development in immunodiagnostics. Applications of this technique in biology and medicine are also highlighted. The scope of immunodiagnosis is unlimited. I am sanguine that this compendium shall prove to be an indispensable source of information to biologists, medical practitioners and clinical technicians.

This monograph is published under the *International Series on Biotechnology*, of which I happen to be the editor. Ever-increasing interest in biotechnology and considering development elsewhere in the world, it offers immense opportunities for a better quality of life and increased agricultural productivity. It is, therefore, imperative to create a strong knowledge-base to provide direction to the ongoing and forthcoming studies. Hopefully, the series will serve this purpose.

The science of immunology has been growing rapidly and, therefore, its applications in the form of immunodiagnostics are bound to flourish. I hope this monograph will find its utility in generating awareness and excitement among students and the professionals alike.

S.C. RASTOGI

Acknowledgements

A major part of this monograph was written while I was on the faculty of Birla Institute of Technology and Science, Pilani, for which I owe thanks to the Director of the Institute. To Mr. Karthik Venkatakrishnan, my student of M.Sc.(Hons) biological sciences, I owe special thanks, for he was instrumental in the search of much desirable information on many immunodiagnostic kits. I am grateful to many scientists on whose data and concepts I have depended on, and who have really shaped this field of immunodiagnostics. I hope that I have fairly and accurately dealt with their contributions.

In a work of this nature, certain commercial systems in common use have been provided as illustrations. For many diseases, there are many manufacturers who produce similar kits, and, therefore, it is difficult to decide which one to use as the illustrative material. All manufacturers had been most cooperative and were willing to supply illustrative material. However, it was always not possible to illustrate alternative example for want of space. Special thanks are due to the British Biocell International for the valuable material on Immunogold labelling provided by them on request. Thanks are also due to Messrs J. Mitra & Bros Pvt. Ltd., Biopool, Canada; Abbott Laboratories, North Chicago; Biocarb Diagnostics, Lund, Sweden; Hycor Biomedical Inc., USA; R&D Systems, Minneapolis; Clonatec, Paris; Dako Diagnostics Ltd., UK; BARC, Bombay; and Span Diagnostics, Udhana, Surat; for providing extensive and exclusive information on their immunodiagnostic kits and reagents.

Lastly, I wish to thank my wife Sarala for showing tremendous patience, and to my son Parag and daughters Smita and Namita, for providing me an inspiring atmosphere in this venture.

S.C. RASTOGI

Contents

Glossary of Acronyms and Abbreviations

ABTS – 2, 2-azinodi-3-ethylbenzthiazoline sulphonic acid
AFP – alpha foetal protein
ANCA – antineutrophil cytoplasmic antibody
CEA – carcinoembryonic antigen
CFT – complement fixation test
DIC – disseminated intravascular coagulation
DNP – dinitrophenyl
DVT – deep venous thrombosis
EIA – enzyme immunoassay
ELISA – enzyme linked immunosorbent assay
EM – electron microscopy
EU – ELISA units
Fab – fragment antibody binding
Fc – fragment crystallisable
FIA – fluoroimmunoassay
GBM – glomerular basement membrane
hCG – human chorionic gonadotropin
HSA – human serum albumin
IF – immunofluorescence
IIF – indirect immunofluorescence
LH – Luteinizing hormone
LHRH – Luteinizing hormone release hormone
ng – nanogram
NTD – neural tube defect

PAN	– polyarteritis nodosa
PBS	– phosphate buffered saline
PEF	– polyethylene glycol
PE	– pulmonary embolism
pg	– picogram
RF	– rheumatoid factor
RIA	– radioimmunoassay
RPCFT	– reiter protein complement fixation test
RPR	– rapid plasma reagin
SRBC	– sheep red blood cells
SRID	– single radial immunodiffusion
STS	– standard test for syphilis
TBG	– thyroxin binding globulins
TfR	– transferrin receptors
THAM	– tris hydroxymethyl amino methane
TPHA	– *Treponema pallidum* haemeagglutination test
TPI	– *Treponema pallidum* immobilisation test
WG	– Wegener's granulomatosis
WR	– Wasserman reaction

CHAPTER 1

The Basis of Immunodiagnostics

Introduction

The world we live in is filled with a gamut of agents that are "foreign" to the human system. Many of them are infectious and have the potential to infect us, that is, they can enter the human body and alter with the normal structure and functions of the body's physiological system. Viruses, bacteria, fungi, protozoa and worms are among the many such infectious agents to which the human body is exposed. Besides, there exist some like dust, pollen, etc., which are capable of causing a state of *allergy.* Well, how does the human body protect itself from these agents which are foreign to it? This question is answered by the presence of an *Immune system* in us whose function is to recognise what is foreign and stage a foray against such agents.

The main role in this process is of a cell called the *lymphocyte.* Immunity is conferred by two types of lymphocytes : the T-lymphocytes and the B-lymphocytes, which constitute two separate but functionally-related limbs of the immune system; namely the *cell mediated* and the *humoral* limbs, respectively. By the term cell mediated, we mean that the role of the cell is central in the process of encountering the foreign agents. In immunological terms, this foreign agent is referred to as 'Antigen' and it is tackled by the T-lymphocyte which is assisted by several other factors, most importantly, the *complement system,* which is a group of self reactive proteins present in the blood plasma.

In the humoral limb of immunity, the B-lymphocytes are involved. When these B cells are stimulated by encountering an outside substance that is antigenic, they produce and release certain proteins which are known as *immunoglobulins.* The encounter itself is a committed process which involves the participation of not only B cells but also of other mononuclear cells in blood. Unlike the cell mediated immunity, in the humoral mechanism, the central role is played by a soluble protein, the immunoglobulin molecule. The most prominent product of the lymphocytes secreted in largest amounts is the immunoglobulin. Immunoglobulins are also known as antibodies because of the interaction of the antigen. With the immunoglobulin *in vivo* results sometimes, but not always, a series of reactions leading to the elimination of the antigen. The antibody molecule acts as an adaptor for the antigen, and when the two encounter each other, either *in vitro* or *in vivo,* the immunoglobulin is capable of undergoing an immunologic reaction in the absence of its cell of origin. This is a bimolecular reaction that follows conventional chemical binding kinetics. The reaction being highly selective and specific, it is of great value for the organism, and can also be exploited in a variety of ways for the purpose of diagnosis as well as therapy of many pathophysiological conditions.

The Immunoglobulin Molecule

The antibody molecule is a four-peptide unit made up of two identical heavy and two identical light chains held together by inter-chain disulphide bonds (Fig. 1.1). These chains can be separated by the reduction of the S-S bonds and acidification. The exposed hinge region is extended in structure due to the high proline content, and can thus be cleaved by proteolytic attack. The action of the enzyme pepsin on the molecule releases two identical fragments, each with a single combining site for antigen (Fab; fragment antigen binding), and a third fragment which cannot bind antigen is designated Fc (fragment crystallizable). Action of the enzyme pepsin, however, cleaves the molecule at a different point and releases the Fc from the rest of the molecule which is divalent with respect to antigen binding like the parent antibody and is hence formulated as $F(ab')_2$. (Fig. 1.1). The antibody population in any given individual is incredibly heterogeneous and the amino acid sequencing of a number of such proteins indicates that the N-terminal portions of both heavy and light chains show considerable variability, whereas the remaining parts of the chains are relatively constant. These variable regions in turn have certain regions with remarkable diversity and are

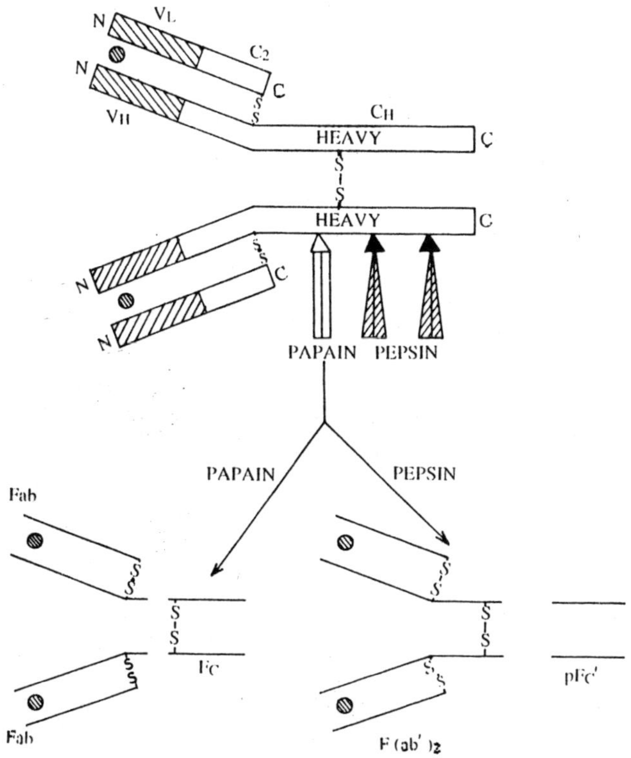

Fig. 1.1. The immunoglobulin molecule and its proteolytic degradation. (• = antigen binding site)

characterised as hypervariable sequences. There are three such hypervariable segments on the variable portions of the heavy as well as the light chains.

Isotypes

Based upon the structure of their heavy chain constant regions, immunoglobulins are classed into major groups termed classes which may be further subdivided into subclasses. In human beings, there are five classes: IgG, IgA, IgM, IgD and IgE. These classes are different not only at the level of the amino acid sequences but also by the antigenic structures to which these sequences give rise. This can be explained by considering myeloma proteins. In the human disease, known as *multiple myeloma,* one cell making one particular individual immunoglobulin divides uncontrollably and gives rise to a clone of cells that produces only one type of immunoglobulin, the myeloma protein or M protein which appears in the urine, sometimes in very high concentrations. By purifying the M protein, we can obtain a preparation of an immunoglobulin having a unique structure. If a human IgG myeloma protein is injected into a rabbit, we can raise an antiserum which can be absorbed by mixtures of myelomas of other classes to remove cross-reacting antibodies. Thus what will be left after this, in the antiserum will be immunoglobulins capable of reacting with IgG, but not with IgA, IgM, IgD or IgE. (Fig. 1.2).[1]

Since all the C_H structures which give rise to classes or subclasses of immunoglobulins are expressed together in the serum of a normal person, they are termed isotypic variants (Table 1.1). Similarly, the C_l regions exist in isotypic forms known as κ and λ which are associated with all heavy chain isotypes. The two light chains in a given antibody molecule being identical, they can both be either of κ type or of λ type. Thus IgG, for example, exists as Ig Gκ or IgG λ, IgM as IgMκ or IgMλ and so on; and the presence of both κ and λ chains in the same immunoglobulin molecule is not seen.

Table 1.1 Summary of immunoglobulin variants

Type of Variation	Distribution	Variant	Location	Examples
Isotypic	All variants present in serum of normal individual	Classes Subclasses Types Subgroups Subgroups	C_H C_H C_L C_L V_H/V_L	IgM, IgE IgA1, IgA2 κ,λ λO_2^{+}, λO_2^{-} VκI, VκII, VκIII. V_{HI}, V_{HII}, V_{HIII}
Allotypic	Alternative forms genetically controlled, so not present in all individuals	Allotypes	Mainly C_H/C_L Sometimes V_H/V_L	Gm groups (humans) b4, b5, b6, b9 (rabbit light chains) Igh-1[a], Igh-1[b] (mouse γ2a heavy chains)
Idiotypic	Individually specific to each immunoglobulin molecule	Idiotypes	Variable regions	Probably one or more hypervariable regions forming the antigen-combining site.

Allotypes

Genetic markers are provided by the existence of allelic forms which essentially refer to the presence of alternative genes at the same locus in

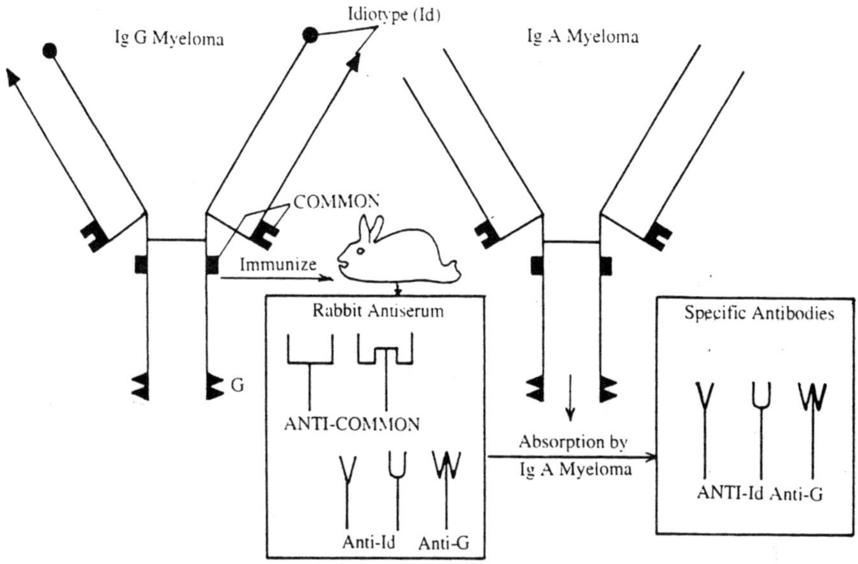

Fig. 1.2. The use of monoclonal myeloma proteins to produce antibodies specific for different Ig structure

the gene regions encoding for the immunoglobulin molecules. Such variants are referred to as *allotypes* and are inherited in a fashion similar to the inheritance of blood group antigens. In human beings, typical allotypes are the Gm specificities on IgG (Gm = marker on IgG). If an individual's IgG has a Gm marker, the test used to check it, is as follows:

Rhesus-D red cells are coated with anti-D IgG bearing the Gm marker we are looking for. These are then mixed with sera from patients with *rheumatoid arthritis* (The disease is characterised by the presence of anti IgG auto-antibodies) containing the appropriate anti-Gm rheumatoid factors, which have been previously incubated with the test IgG. In case the test IgG has the appropriate marker, a positive binding reaction with the rheumatoid factors (RF) would occur in the preincubation, and no binding sites on the RF molecule will be available for reaction with the anti-D's Gm markers upon mixing. Thus, no agglutination of the red cells will occur (Fig 1.3).[1] Allotypic differences at Gm locus usually involve one or two amino acids in the peptide chain. The G1m (a) locus on the IgG1, for example, would have the peptide sequence Asp-Glu-Leu-Thr-Lys. If the IgG1 is a^- negative, the sequence would be Glu-Glu-Met-Thr-Lys, i.e. two amino acids are different.

Totally, about 25 Gm groups have been found on the γ-heavy chains and a further three (the Km-previously Inv-groups) on the *k* constant region.

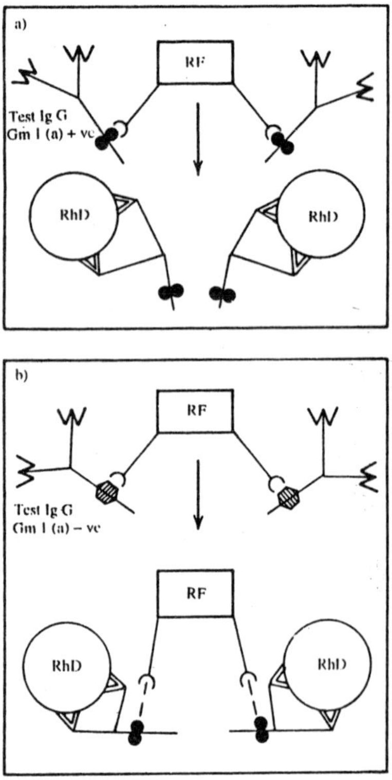

Fig. 1.3. Determination of allotypic markers on IgG
(a) No agglutination of redcells when test IgG is Gm1(a) +ve
(b) Agglutination of red cells when test IgG is Gm1 (a) –ve

Allotypic markers have also been found on the immunoglobulins of rabbits and mice using reagents prepared by immunising one animal with an immune complex obtained with the antibodies of another animal of the same species. The genes encoding these markers are expressed co-dominantly and inherited in a simple Mendelian fashion. For example, in rabbit light chains, b4, b5, b6, b9 are the allotypic variants.

A rabbit of $b^4 b^5$ genotype derived from $b^4 b^4$ and $b^5 b^5$ parents would express the b_4 marker on one fraction and the b_5 on another fraction of its immunoglobulin molecules.

Idiotypes

Immunoglobulins, which are isotypically as well as allotypically similar, may still be different and are referred to as idiotypes. It is possible to raise antisera which are specific for individual antibody molecules and discriminate between one monoclnal antibody and another independent of Isotypic or allotypic structures. Such antisera in fact define the individual determinants characteristic of each antibody, collectively called the *idiotype.*

These idiotypic determinants are located in the variable part of the antibody associated with the hypervariable regions. Idiotypes are of two types: private and public or cross-reacting. These are explained by the reaction of the immunoglobulins with anti-idiotypic sera. Anti-idiotypes which react only with one antibody are said to recongnise private idiotypes. These give further support to the idea that each antibody has a unique structure. The other type of idiotypes, namely the public or the cross-reacting idiotypes, as their names indicate, are those shared by many antibodies of closely similar amino acid structure.

Anti-idiotypic sera provide useful reagents for demonstrating the same V region on different heavy chains and on different cells, for the identification of specific immune complexes in patients' sera, for selecting lymphocytes with certain surface receptors and for a variety of other diagnostic and prognostic purposes. The main application of anti-idiotypic sera is in the diagnosis and prognosis of multiple myeloma. In this disease, the excess light chains of immunoglubulins which are produced monoclonally, are released in the urine as homogeneous, free light chains. Being identical, the variable region fragments polymerise and form the characteristic amyloid fibrils seen in the urine. Recognition of V_L type amyloid in subjects excreting *Bence Jones Protein* (the light chains appear as the Bence Jones Protein) is possible by the use of anti-idiotypic sera. Also detection of residual monoclonal protein after therapy is possible by the use of anti-idiotypic antisera.

Binding of Immunoglobulins to Antigen

As explained earlier, any substance that is foreign to the system, when enters it and stimulates the immune system to produce specific antibodies, the immunoglobulins produced will specifically bind it to the foreign substasnce or 'antigen' and what results in an antigen-antibody reaction. In the Y-shaped structure of the immunoglobulin molecule, the parts of the molecule which bind to the antigen are the variable regions, namely V_L and V_H which form the terminal part of the Y-shaped molecule. The basic structure described is thus divalent with respect to the antigen.

Immunoglobulins are not simple linear sequences of amino acids in the shape of a Y. In addition to the inter-chain disulphide bonds, there are intrachain disulphide links which form loops in the peptide chain. These loops actually exist in a compactly folded form giving rise to globular domains. The hypervariable regions which are very important for antigen recognition and the binding appear at one end of the variable domain where they form parts of the β-turn loops of the β-pleated structure, thereby remaining clustered close to each other in space. This has been indicated by X-ray crystallorgraphic studies of a Bence Jones Protein by Schiffler *et al.* (1973).[2]

The fact that the hypervarible regions are the regions of the antibody molecule which bind to the antigen is evident from the clustering of the hypervariable loops at the tips of the variable regions where the antigen binding site is located. This has also been confirmed by X-ray crystallographic studies. The binding of immunoglobulin to antigen is shown in Fig. 1.4. Each hypervariable region in the cluster is complementary to a certain specific part of the antigen determinant. The sequence heterogeneity of three heavy chains and three light chains hypervariable regions ensures tremendous diversity in combining specificity for antigen through variation in the shape and nature of surface they create.

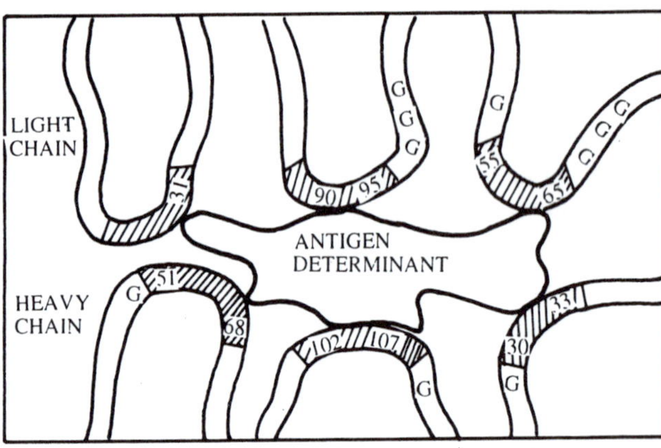

Fig. 1.4. Idealized two-dimensional representation of an (G = Glycine) antigen binding site formed by spatial apposition of peptide loops containing the hypervariable regions.

A *hapten* is a substance that is not immunogenic but can react with a specific antibody. An *epitope* is the simplest unit in an antigen or hapten capable of reacting with antibody, while the simplest unit in the antibody that combines with the antigen is called *paratope*. Affinity-labelling is a technique which can be used to identify the amino acids that are associated with the antigen binding site. In this technique, a hapten (e.g.,DNP*) is equipped with a chemically reactive side chain which will form covalent links with adjacent amino acids after combination of the hapten with the

* DNP – Dinitrophenyl.

antibody, thereby labelling residues in the neighbourhood of the combining site. Porter and his colleagues utililsed a 'flick-knife' principle as a modification of this method. The hapten with an azide side chain combines with its antibody and is then illuminated with ultraviolet light; this converts the azide group into the reactive nitrene radical which will covalently link to almost any organic group with which it comes in contact. It has been seen tht the affinity label binds both the heavy and light chains in the hypervariable regions. An example of this method is shown in Fig. 1.5. As evident from the figure, the amino acid residue which has got labelled in this hypothetical example is alanine. Analysis of the protein after digestion would show alanine to be labelled with the hapten and implicate this residue in the combing site[3].

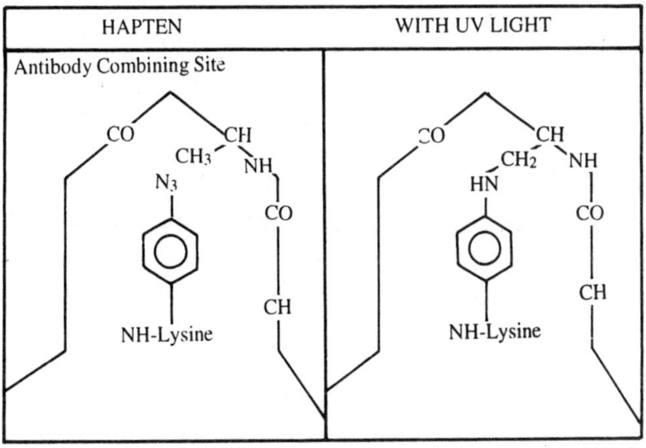

Fig. 1.5. Illustration of flick-knife principle of affinity labelling

The Complement System

Complement is the name given to a complex series of some 20 proteins which, along with blood clotting, fibrinolysis and kinin formation, forms one of the triggered enzyme systems found in the plasma. These systems characteristically produce a rapid, highly amplified response to a trigger stimulus, mediated by a cascade phenomenon where the product of one reaction is the enzymic catalyst of the next. Some of the complement components are designated by the letter 'c' followed by a number which is related more to the chronology of its discovery than to its position in the reaction sequence. The most abundant and most important component is C3 which has a molecular weight of 195 KD and is present in plasma at a concentration of around 1.2 mg/ml.

Either due to reaction with water or due to the reaction with the trace amounts of plasma proteolytic enzyme, the internal thioester bond in C3 becomes activated and the molecule splits to form a larger product C3b and a smaller one C3a (In cleavages of a complement component, the larger product is generally given the suffix 'b' and the smaller 'a'). In the presence of M_g^{2+}, the C3b can complex with another Complement component, factor B, which then undergoes cleavage by a normal plasma enzyme (factor D) to generate $\overline{C3bBb}$ (the bar over the complex denotes its enzymic activity). This complex is known as 'C3 convertase' and can by virtue of its enzyme activity split C3 to give C3a and C3b. Thus the process is cyclic as shown in Fig. 1.6. C3 cleavage has important biological consequences in relation to microbial defenses. In the absence of microbial activation, that is under normal conditions; in solution, the C3bBb convertase is unstable and factor B is readily displaced by another component factor H. This forms a complex that is susceptible to attack by C3b inactivator (factor I). The inactivated C3b is biologically inactive and undergoes further degradation by typsin-like enzymes in the body fluids.

A number of microbes can activate the $\overline{C3bBb}$ convertase to generate large amounts of C3 cleavage products by stabilising the enzyme on the carbohydrates on their surfaces, thereby protecting the C3b from factor H. Further stabilisation of the bound convertase occurs by another protein, *properdin*. The C3 convertase acts on C3, and as the C3 is split by the surface membrane-bound enzyme, its internal thioester bond becomes activated and the reactive intermediate C3b binds covalently to the membrane. Thus the microbial membrane is now deposited with C3b molecules. This series of reactions leading to C3 breakdown provoked directly by microbes is called 'The Alternative pathway' of complement activation. Although in the phylogenetic terms, this is the oldest pathway, it was discovered after a separate pathway and so is designated as alternative. The other pathway of complement activation is known as the 'Classical Pathway' of complement activation. This pathway is antibody dependent. In fact, it is the antibody which initiates the classical complement sequence. In this pathway, antibody, when bound to a microbe, will link to and activate the molecule C1q, the first in the sequence. This molecule is polyvalent with respect to antibody binding and consists of a central collagen-like stem branching into six peptide chains, each tipped by an antibody binding subunit. C1q is associated with two further subunits, C1r and C1s, in a Ca^{2+} stabilised trimolecular complex. These are activated sequentially and the complex now attains proteolytic activity and acts on C4 breaking it into $\overline{C4b}$ and C4a. $\overline{C4b}$ now binds either to the antibody-C1 complex or the surface of the microbe itself. In the presence of Mg^{2+}, C2 can complex with $\overline{C4b}$ to form C4b2 which again gets broken into C4b2b by C1qrs. The

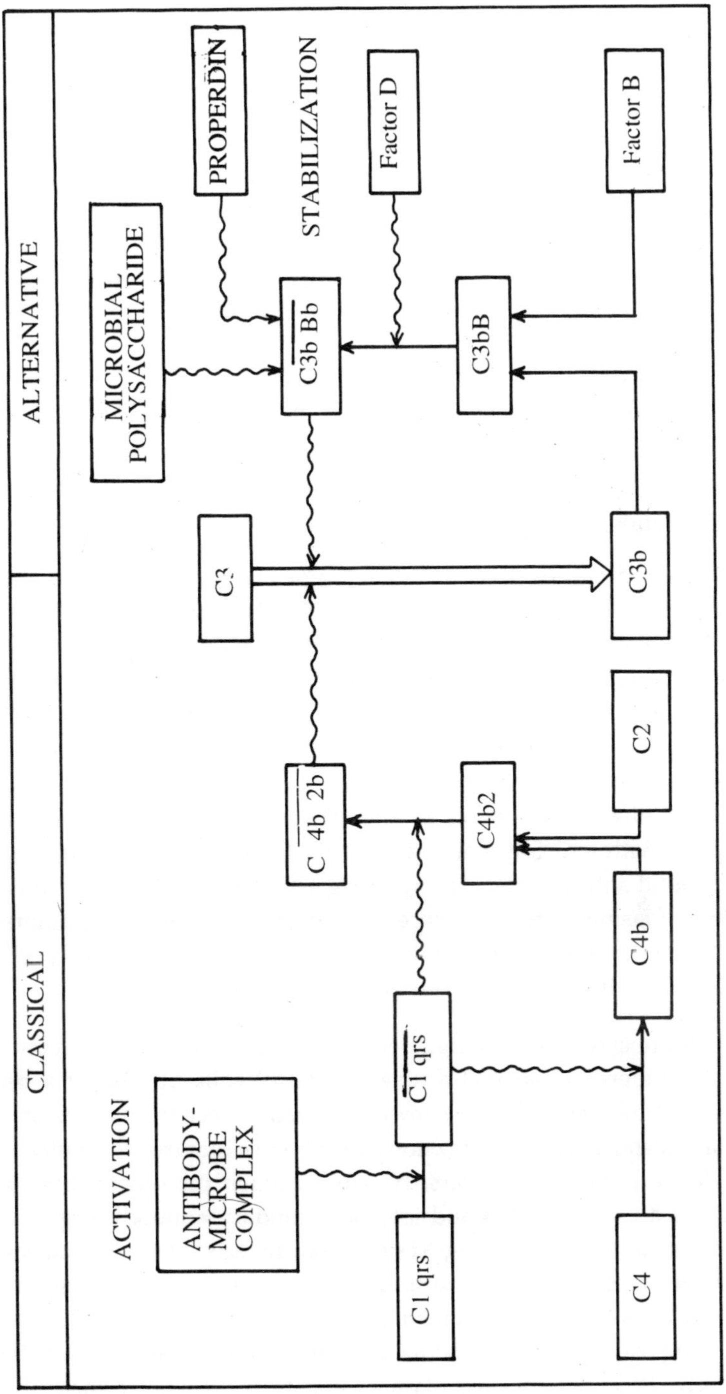

Fig. 1.6. Comparison of the classical and alternative pathways of complement

resulting $\overline{\text{C4b2b}}$ has the C3 convertase activity required to activate C3 with the same specificity as the $\overline{\text{C3bBb}}$ generated by the alternative pathway (Fig. 1.6).

The post - C3 pathway is parallel for both the classical and the alternative pathways and involves the activation of C5. C5 binds to the membrance bound C3b. Then, it becomes a substrate for $\overline{\text{C3bBb}}$ and is cleaved to release a small polypeptide C5a, leaving the larger fragment, C5b on the membrane. This binds C6, C7, and then C8 to form a complex which induces a critical orientation between two or more molecules of the terminal component C9. This leads to unfoliding of C9, insertion into the lipid bilayer and polymerisation to an annular 'membrane attack complex' (MAC) forming a transmembrane channel fully permeable to electrolytes and water. Due to high internal colloid osmotic pressure of cells, there is a net influx of Na^+ and water frequently leading to lysis.

Immunoglobulin Classes

As discussed earlier, there are five major classes of immunoglobulins, namely IgG, IgA, IgM, IgD, and IgE. The IgG molecule and its basic structure is given in Fig 1.1. Its valency for binding antigen is 2 and the heavy chains are of γ type. Both κ and λ light chains are possible. When an animal is given an injection of an antigen, the antigen selects the lymphocytes which make the antibody. The antigen is confronted with a dazzling array of lymphocytes, all bearing different antibodies, each with its own individual recognition site. The antigen will bind only those receptors with which it makes a good fit. Lymphocytes whose receptors have bound antigen receive a triggering signal and develop into antibody forming plasma cells, and since the lymphocytes are programmed to make only one kind of antibodies, those secreted by the plasma cell will be identical with those originally acting as lymphocyte receptors.

Lymphocytes are thus triggered by exposure to antigen, then undergo clonal proliferation so that a single clone of cells producing a certain antibody develops. This is known as clonal selection and results in a primary response, producing enough antibodies required to combat infection. If the animal encounters the same antigen the second time, what results is a much quicker and intense secondary antibody response because the antibody forming system has already been established by previous exposure and only a tuning up is necessary. In this secondary response, it is IgG which is the major immunoglobulin to be synthesized. The fixing of complement described earlier can be brought about by IgG as antibody. IgG can cross the placenta and thus provides a major line of defense against infection for the first few weeks of a baby's life. Ig G can neutralize bacterial toxins and binds to microorganisms to enchance their phagocytosis, particularly in internal body fluids, mainly in the extravascular region. IgG is the only antibody that can cross placenta.

IgA appears selectively in the sero-mucous secretions such as saliva, tears, nasal fluids, sweat, colostrum, and secretions of the lungs, genitourinary and gastrointestinal tracts where it defends the exposed external surfaces of the body against an attack by microorganisms. IgA is present in these fluids as a dimer stablised against proteolysis by combination with another protein — the secretory component which is synthesised by the local epithelial cells. Dimerisation of IgA occurs within the plasma cells where it is synthesised. The dimerisation occurs with a cysteine-rich polypeptide called J chain of molecular weight 15,000.

IgM is often referred to as macroglobulin antibody. IgM molecules are polymers of five four-peptide subunits, each bearing an extra C_H domain. As in IgA, the J-chains are necessary for polymerisation. The molecule has a star-shape and a theoretical antigen binding valency of 10, which of course, is observed only when it interacts with small haptens. With larger antigens, the valencey falls to 5. Due to their high valency, these antibodies are extremely efficient agglutinating and cytolytic agents. They are produced early in the immune response due to infection and form an effective first line defence against bacteraemia. The isohemagglutinins and many of the 'natural' antibodies to microorganisms are usually IgM; antibodies to the typhoid 'O' antigen (endotoxin) and the 'WR' antibodies in syphilis also tend to be found in this class. Monomeric IgM with a hydrophobic sequence linked to the C-terminal of the heavy chains to anchor the molecule in the cell membrane, is the major antibody receptor used by B-lymphocytes to recognise antigen.

IgD is different from the other classes of immunoglobulins in that it has an extended hinge region which though protected by carbohydrate, makes the molecule more prone to proteolysis when compared to other classes of antibodies. This accounts for its short half-life in blood plasma, (of 2.8 days). Almost all IgD is present, together with IgM on the surface of a portion of lymphocytes. This could mean that they operate as mutually interacting antigen receptors for the control of lymphocyte activation and suppression.

IgE is present in very low concentration in the serum and is produced only by a very small fraction of body's lymphocytes. IgE antibodies remain firmly fixed for an extended period when injected into human skin where they are bound to mast cells. Contact with antigen leads to degranulation of the mast cells with release of vasoactive amines. This process is responsbile for symptoms of hayfever and extrinsic asthma when patients with atopic allergy come into contact with the allergen. The main physiological role of IgE is the protection of external mucosal surfaces of the body through an acute inflammatory reaction.

The physical properties of the major immunogloblin classes are listed in Table 1.2. Apart from these classes, further subdivision into isotypic subclasses has been seen. IgG can have either of the 4 types of heavy chains, γ_1, γ_2, γ_3, and γ_4. Based on this, we have IgG1, IgG2, IgG3 and IgG4 respectively. Similarly IgA1 and Ig A2 are also seen.

Table 1.2 Physical properties of major human immunoglobulin classes

Designation	IgG	IgA	IgM	IgD	IgE
Sedimentation Coefficient	7S	7S,9S,11S	19S	7S	8S
Molecular Weight	150,000	160,000 and dimer	900,000	185,000	200,00
Number of basic four-peptide units	1	1,2*	5	1	1
Heavy Chains	γ	α	μ	δ	ε
Light Chains	$\kappa+\lambda$	$\kappa+\lambda$	$\kappa+\lambda$	$\kappa+\lambda$	$\kappa+\lambda$
Molecular formula	$\gamma_2\,\kappa_2$, $\gamma_2\,\lambda_2$	$(\alpha_2\,\kappa_2)_{1-2}$ $(\alpha_2\,\lambda)_{1-2}$ $(\alpha_2\,K_2)_2\,S*$ $(\alpha_2\,\lambda_2)_2\,S*$	$(\mu_2\,\kappa_2)_5$ $(\mu_2\,\lambda_2)_5$	$\delta_2\,\kappa_2$ $(\delta_2\,\lambda_2\,?)$	$\varepsilon_2\,\kappa_2$, $\varepsilon_2\,\lambda_2$
Valency for Antigen binding	2	2,4	5(10)	2	2
Concentration range in Normal serum	8-16 mg/ml	1.4-4 mg/ml	0.5-2 mg/ml	0-0.4- mg/ml	17-450 ng/ml
% Total Immuno-globulin	80	13	6	0–1	0.002
% Carbo-hydrate-content	3	8	12	13	12

* Dimer in external secretions carries secretory component-S

Immunodiagnostics : Importance and Scope

In many pathological conditions, certain unusual proteins or antigens appear in certain body parts or in circulation. Certain endocrine disorders may, for example, be characterised by decreased or increased levels of certain hormones. Hypothyroidism, for example, is characterised by decreased levels of thyroxine. Another example is that of tumor markers. Hepatocellular carcinoma, for example, is characterised by abnormally elevated levels of Alpha Foeto Protein in the blood. In all the above mentioned cases, the presence of or the quantity of the normal or unsual substances in the body can be done by detecting the reaction of that substance with its complementary antibody. Such assays employing immune substances, namely antibodies either for qualitative, semi-quantitative or quantitative analyses are broadly termed as immunoassays. In other words, immunoassays are methods, utilising immunological reactions for the measurement of substances of biological or clinical interest. Such assays have inherent exquisite specificity due to their immunological nature, and by various procedures can have extremely high sensitivity. In the past, immunoassays were relatively simple (e.g., as in blood grouping and in bacterial agglutination tests), but the increasing appreciation of their versatilty has led to many recent highly complex assays. In the beginning, the tests were almost done manually but now these are being replaced by automated systems often adapted to large scale batch processing of samples. At the same time, progress in electronics has led to vastly improved data-processing capabilities on instruments allowing much better interpretation of the results of the tests. Immunodiagnostics has widespread applications in the fields of blood banking, endocrinology, cancer diagnosis and prognosis, autoimmunediseases, genetic diseases, veterinary science, microbiology and virology.

When an antigen-antibody reaction occurs, it leads to the formation of complexes which can be visualised macroscopically. This forms the basis of immunodiagnosis. Highly specific monoclonal antibodies can be produced by the hybridoma technique and offer high specificity in such assays. Apart from the direct methods of visualising antigens/antibodies, there are other methods in which the labelled reagents are used. Such methods are better used in diagnosis though high cost and sophisticated instrumentation limit their use to specific cases. The major immunodiagnostic techniques are classified as follows.

IMMUNODIAGNOSIS

DIRECT METHODS | LABELLED METHODS

1. Precipitation
2. Agglutination
3. Neutralization
4. Complement Fixation Tests

1. Enzyme Immunoassays
2. Immunofluorescence
3. Radioimmunoassays

Beside these techniques, other special areas of pathology where immunology plays a major role include *in vivo* allergy testing, human leucocyte antigen typing and immunosensing methods and their applications in cell sorting.

The field of immunodiagnostics is very vast and its applications are widespread. The aim here will be to study the principles and methodology of the various immunoassays presently used for the diagnosis of various infections as well as noninfectious diseases. The emphasis will be more on the recent findings and the actual application in the serological laboratory.

CHAPTER 2

The Antigen-Antibody Reaction

Introduction

Once antibodies were raised to small chemically defined haptens, it became possible to relate variations in the chemical structure of a hapten to its ability to bind a given antibody. In one experiment, antibodies raised to m-aminobenzene sulphonate were tested for their ability to combine with ortho, meta and para isomers of the hapten and related molecules in which the sulphonate group was substituted by arsenate or carboxylate. (Table 2.1).[4] The hapten with the sulphonate group in the ortho position combines somewhat less well with the antibody than the original meta-isomer, but the para substituted compound which, though chemically similar to the ortho, shows very poor reactivity. The substitution of arsenate for sulphonate leads to weaker combination with the antibody; both the groups are negatively charged and have a tetrahedral structure but the arsenate group is larger in size and has an extra H atom. The aminobenzoates in which the sulphonate group is replaced by the negatively charged but planer carboxylate group show even less affinity for the antibody. So, we can say that the overall configuration of the hapten is even more important than its chemical nature, i.e., the hapten is recognised by the overall three-dimensional shape of its electron cloud as distinct from its chemical reactivity. Thus the antibody may be directed against a particular electron-cloud shape rather than a specific chemical structure. This view is consistent with the nature of antigen-antibody binding which is known not to involve covalent linkages and, like enzyme-substrate interactions, can be readily reversed. The antibody actually recognises antigen through complementariness of shapes on paratope and epitope as confirmed by x-ray crystallographic analysis of the interaction of the Fab fragment of monoclonal anti-lysozyme with its antigen.

Table 2.1 Effect of variations in hapten structure on strength of binding to antibodies raised against m-aminobenzene sulphate

	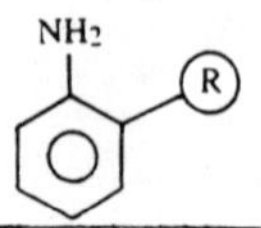 ORTHO	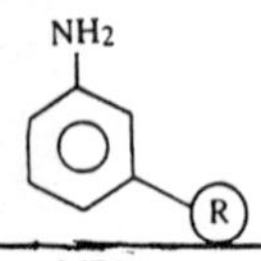 META	 PARA
R = Sulphonate 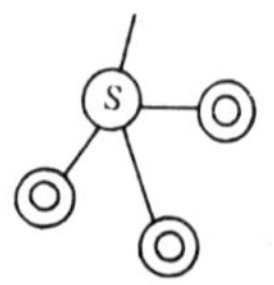 tetrahedral	++	+++	±
R = Arsonate 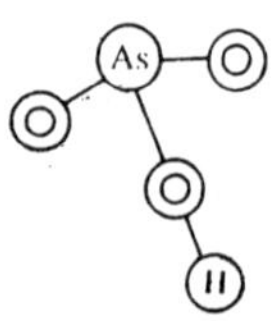 tetrahedral	–	+	–
R = Carboxylate 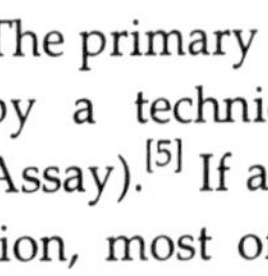 Planar	–	±	–

Antigen-Antibody Binding

The primary binding between the antigen and antibody can be measured by a technique known as ammonium sulphate precipitation (Farr Assay).[5] If ammonium sulphate is added to dilute serum to 50% saturation, most of the immunolglobulin is precipitated, while other serum proteins such as albumins remain in solution. This is the basis of the Farr assay. The test, in its original form is only suitable for antigens which are soluble in 50% saturated ammonium sulphate solution.

Antigen, often albumin, is radiolabelled and allowed to react with antibody, generally in antigen excess, so that soluble complexes are formed. An equal volume of saturated ammonium sulphate is added and all the immunoglobulin is precipitated. Only the antigen complexed with antibody is precipitated under these conditions. The precipitates are washed with 50% saturated ammonium sulphate solution to remove any free antigen and then counted for radioactivity. The amount of radioactivity in the precipitate is proportional to the amount of antigen bound by the antibody and the results are expressed in terms of antigen binding capacity per ml of serum. For accurate determinations in research procedures, the results need to be calculated to the same degree of antigen excess for each antiserum. Therefore, in the original assay, constant amounts of antigen are added to a series of dilutions of serum and the antigen binding capacity determined for each dilution.

Figure 2.1 shows a typical binding curve. It has become accepted that all the results are calculated on the 33% end point, i.e., the dilution of serum which binds 33% of added antigen.

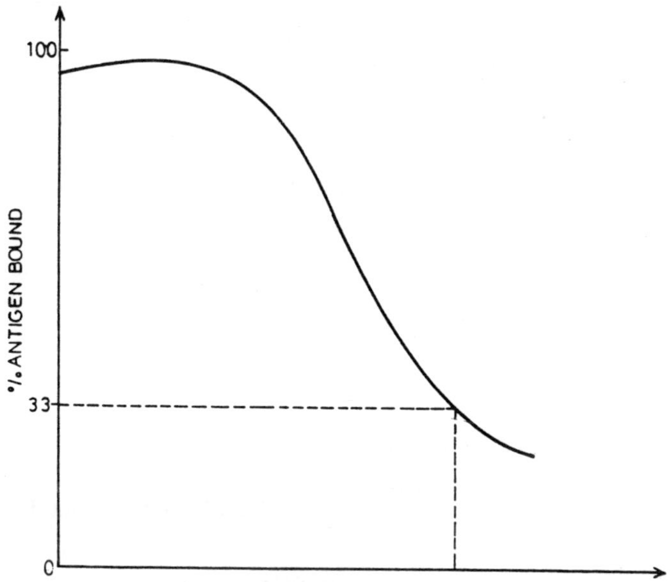

Fig. 2.1. Antigen binding capacity of antiserum determined by Farr assay

So far, we have considered antibody-antigen reaction in terms of the quantity of antibody produced. For each antigen determinant, a range of different antibodies are formed, some of which will fit the antigen more strongly than the poorer fitting molecules. This leads to the concept of antibody affinity. Affinity is a measure of the concept of antibody affinity. Affinity is a measure of the strength of antigen-antibody combination. Affinity actually refers to the interaction between multivalent antigenic determinants, i.e., haptens and single antibody combining sites. When dealing with multivalent antigens and antibodies, the interaction is more complex and the term *avidity* is used.

Kinetics of Antigen-Antibody Reaction

A technique, based on the Farr assay, has been developed which is suitable for affinity or avidity determinations of unselected and unpurified antibodies. The combination of antigen (Ag) and antibody (Ab), to form a complex is a reversible reaction :

$$\text{Ab} + \text{Ag} \underset{K_d}{\overset{K_a}{\rightleftarrows}} \text{Ab} \cdot \text{Ag},$$

Where K_a and K_d are the association and dissociation constants, respectively.

Though this is a heterogeneous system, we can approach it by the application of the law of mass action and the reaction affinity is given by the equilibrium constant K (litres mole^{-1}).

$$K = \frac{K_a}{K_d} = \frac{[\text{Ab} \cdot \text{Ag}]}{[\text{Ag}]\ [\text{Ab}]}$$

Ag refers to monovalent antigenic determinants and Ab refers to the independent antibody combining sites.

If K is determined with respect to total antibody binding sites, Ab_t, then we can derive the following form of the Langmuir Adsorption Isotherm.

$$\text{Ab} + \text{Ag} \rightleftarrows \text{Ab} \cdot \text{Ag}.$$

By the law of mass action,

$$\frac{[\text{Ab} \cdot \text{Ag}]}{[\text{Ab}]\ [\text{Ag}]} = K \text{ (equilibrium constant)}$$

therefore,
$$[\text{Ab} \cdot \text{Ag}] = K\,[\text{Ab}]\,[\text{Ag}] \tag{1}$$

Making $[\text{Ab}_t] = [\text{Ab} \cdot \text{Ag}] + [\text{Ab}]$

therefore,
$$[\text{Ab}] = [\text{Ab}_t] - [\text{Ab} \cdot \text{Ag}] \tag{2}$$

Substituting (2) in (1),

$$[\text{Ab} \cdot \text{Ag}] = K\,[\text{Ag}]\,([\text{Ab}]_t - [\text{Ab} \cdot \text{Ag}])$$

Therefore,
$$[\text{Ab} \cdot \text{Ag}] = K\,[\text{Ag}]\,[\text{Ab}_t] - K\,[\text{Ag}]\,[\text{Ab} \cdot \text{Ag}]$$

or
$$[\text{Ab} \cdot \text{Ag}] + K\,[\text{Ag}]\,[\text{Ab} \cdot \text{Ag}] = K\,[\text{Ag}]\,[\text{Ab}_t]$$

or
$$[\text{Ab} \cdot \text{Ag}]\,(1 + K\,[\text{Ag}]) = K\,[\text{Ag}]\,[\text{Ab}_t]$$

Therefore,
$$\frac{[\text{Ab} \cdot \text{Ag}]}{[\text{Ab}_t]} = \frac{K\,[\text{Ag}]}{1 + K\,[\text{Ag}]} \tag{3}$$

Making [Ab · Ag] = b, then (3) may be re-written as

$$\frac{b}{[Ab_t]} = \frac{K\,[Ag]}{1 + K\,[Ag]}$$

or
$$\frac{b\,(1 + K\,[Ag])}{[Ab_t]\;K\,[Ag]} = 1$$

or
$$\frac{1}{b} = \frac{1 + K\,[Ag]}{K\,[Ab_t]\;[Ag]}$$

$$\frac{1}{b} = \frac{1}{K\,[Ab_t]\;[Ag]} + \frac{1}{[Ab_t]} \qquad (4)$$

When $\frac{1}{[Ag]} = 0$, in (4), $\frac{1}{b} = \frac{1}{[Ab_t]}$

As b = [Ab · Ag] = bound Ag concentration, and [Ag] = free antigen concentration, then a plot of $\frac{1}{b}$ against $\frac{1}{[Ag]}$ can be extrapolated to obtain $[Ab_t]$. This is the total of antibody combining sites. It should be noted that the concentrations expressed in the above equations are molar equivalents, hence, for example, when [Ag · Ab] = bound antigen concentration, this is a molar equivalence and not a weight equivalence.

Consider again the mass action equation

$$K = \frac{[Ab \cdot Ag]}{[Ab]\,[Ag]}$$

If increasing amounts of antigen are reacted with a fixed amount of antibody, a point is reached where half the antibody combining sites are occupied by the antigen.

At this point,

$$[Ab] = [Ab \cdot Ag]$$

therefore in the mass action equation, $K = \frac{[Ab \cdot Ag]}{[Ab]\,[Ag]}$

$$K = \frac{1}{[Ag]}$$

In other words, the affinity equals the reciprocal of the free antigen concentration when half the antibody sites are occupied by antigen. The value of *K* can be obtained from the plot of $\frac{1}{b}$ against $\frac{1}{[Ag]}$ by calculating the value of half $[Ab_t]$ and reading off from the graph the value of *K*. (Fig. 2.2).

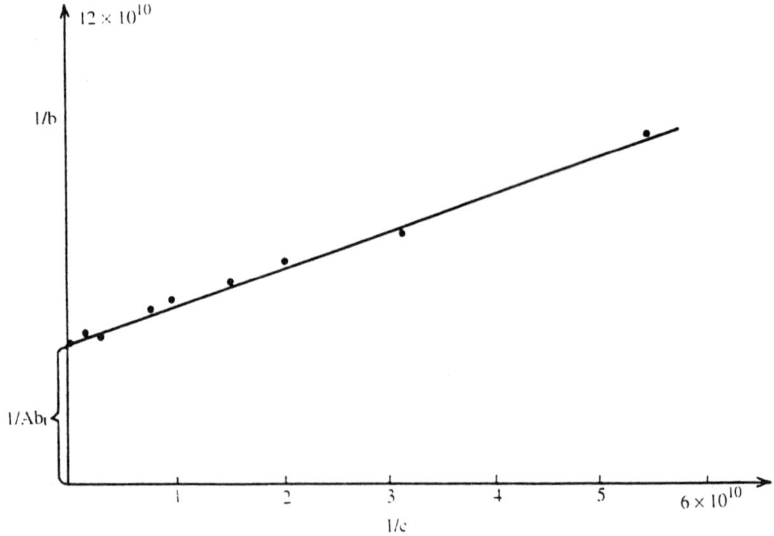

Fig. 2.2. Langmuir plot of reciprocal of bound (1/b) versus reciprocal of free antigen (1/c or 1/Ag), from a baboon immunized with human chorionic gonadotropin (After *Practical Immunology* by Hudson, L. and F.C. Hay, Blackwell (1980))

In this particular experiment, the value for Ab_t obtained by extrapolation to infinite antigen concentration is 3.21×10^{-11} ml^{-1}. Antibody affinity $K = 4.3 \times 10^{10}$ ml^{-1}.

In experimental terms, this value can also be obtained from a plot of the logarithmic transformations of the Sip's equation as follows:

If $$\frac{r}{n} = \frac{(K\,[Ag])^a}{1+(K\,[Ag])^a}$$ where

r = moles of antigen bound per mole of antibody

n = Antibody valency

a = Heterogeneity index

The heterogeneity index, a, is a measure of the number of different molecular species of antibody. It can range from a value of 0 to 1. Low values represent a large value of heterogeniety, whereas a monoclonal antibody should theoretically have an index of 1.

From the Sip's equation,

$$\log \frac{r}{n-r} = a \log K + a \log [A] \qquad (5)$$

for IgG, n = 2

Therefore moles of antibody = $\frac{[Ab_t]}{2}$

and $$r = \frac{b}{[Ab_t]/2} \text{ where } b = [Ab.Ag]$$

Substituting for n and r in (5),

$$\log \frac{b}{[Ab_t] - b} = a \log K + a \log [Ag]$$

If $\log \frac{b}{[Ab_t] - b}$ is plotted against log [Ag], when

$$\log \frac{b}{[Ab_t] - b} = 0,\ K = \frac{1}{[Ag]}.$$

In actual experiment, a constant amount of antibody is reacted with increasing concentrations of antigen. After equilibration, the concentration of bound antigen is detected in the complex after ammonium sulphate precipitation, and the concentration of free antigen is determined in the supernatant. If the reciprocal of bound antigen is plotted against the reciprocal of the free antigen concentration and the line extrapolated to $\frac{1}{[Ag]} = 0$, then the reciprocal of the bound antigen will equal the reciprocal of the total antibody combining sites.

While of term affinity describes the binding of antibody to a monovalent hapten or single antigen determinant, in most practical circumstances we are concerned with the interaction of an antiserum with a multivalent antigen. As mentioned earlier, the term employed to express this binding is avidity.

$$n\,Ab + m\,Ag \longrightarrow Ab_n\,Ag_m$$

The factors which contribute to avidity are complicated, including as they do the heterogeneity of antibodies in a given serum which are directed against each determinant on the antigen, and the heterogeneity of the determinants themselves. Figure 2.3 explains the heterogeneity of IgG anti-hapten (dinitrophenyl: DNP) antibodies from the serum of an immunized animal contrasting with the homogeneity of a monoclonal IgG anti-DNP.

From the law of mass action,

$$K = \frac{[Ab\ H]}{[Ab]\ [H]} \text{ for the reaction}$$

$$Ab + H \rightleftharpoons Ab\,H$$

If r represents the average number of DNP molecules bound to each antibody molecule, of affinity constant K and number of binding sites n, in the presence of a free hapten concentration [H], then from the mass-action equation of equilibrium relationships,

$$K = \frac{r}{(n - r)\ [H]}$$

or

$$\boxed{\frac{r}{[H]} = n\,K - rK}$$

In Fig 2.3a, r has been plotted against $\frac{r}{[H]}$. This is a scatchard plot of hapten binding to antibodies purified from the serum. Thus a scatchard plot of $\frac{r}{[H]}$ against r for a single antibody species will be a straight line of slope K as seen for the monoclonal antibody. The deviation from the straight line given by anti-DNP from the antiserum clearly indicates the existence of antibodies with different affinities as may be confirmed by the binding of labelled DNP to many different bands after the separation of the individual antibodies by isoelctric focussing of the serum. Extrapolation to $\frac{r}{[H]}$ (at infinitely high concentration of antigen) gives the number of binding sites on each IgG molecule as 2. For IgM antibodies, the value would be 10.

Figure 2.3 (b) is a histogram showing a typical distribution of antibody affinities in the anti-DNP serum. Affinities between 10^4 and 10^{10} are measurable. The monoclonal antibody of course gives a single affinity value.[6]

The multivalence of most antigens leads to an interesting 'bonus' effect in which the binding of two antigen molecules by antibody is always greater, usually many-fold greater, than the arithmetic sum of the individual antibody links. This is illustrated in Fig. 2.4. The force binding the two antigen molecules in (c) with two antibody bridges is many-fold greater than (a+b). The binding of the two antibodies to determinants 1 and 2 on the antigen can each be described by the individual change in free energy state on forming the epitope-paratope bond with its corresponding affinity constant, that is

$$\Delta G_1 = -RT\ \log_e K_1 \text{ and } \Delta G_2 = -RT\ \log_e K_2$$

For both antibodies operating in conjunction, the overall free energy change giving the avidity (K_{avid}) would be

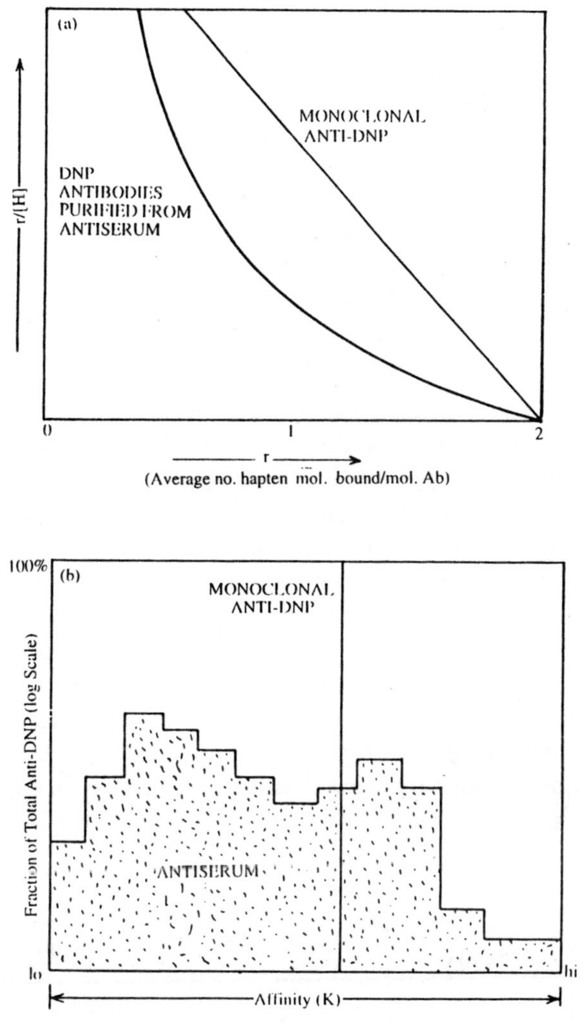

Fig. 2.3. (a) Scatchard plot
(b) Affinity distribution

$$\Delta G = \Delta G_1 + \Delta G_2 = -RT \log_e K_1 - RT \log_e K_2$$

$$= -RT (\log_e K_1 + \log_e K_2)$$

$$= -RT (\log_e K_1 \times K_2)$$

Since $\Delta G = -RT \log_e K_{avid}$

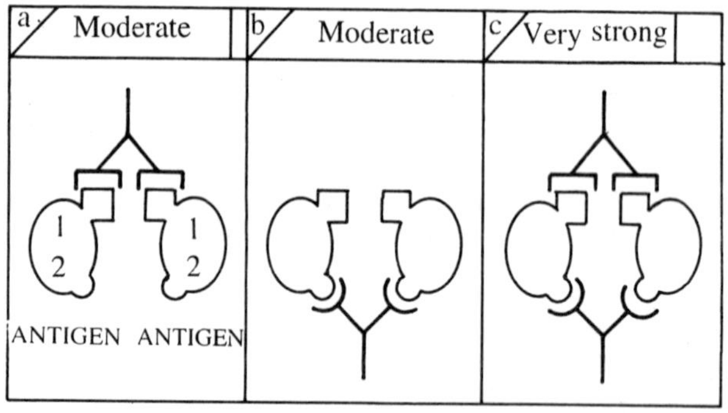

Fig. 2.4. The Bonus Effect of multivalant attachement on binding strength

$$K_{avid} = K_1 \times K_2$$

The tremendous increase in equilibrium constant resulting from multiplying the contributing affinities is responsible for the Bonus Effect. To given a concrete example, if K_1 is 10^4 and K_2 10^3, K_{avid} would be 10^7 l/mole. Figure 2.5 explains the mechanism of the Bonus effect. Each antigen-antibody bond is reversible and with a single antibody bridge between two antigen molecules (a), dissociation of either bond could enable an antigen molecule to 'escape' as in (b). If there are two antibody bridges, even when one dissociates, the other prevents the antigen molecule from escaping and holds it in position ready to reform the broken bond. The same considerations apply to the binding of antibody to a polymeric antigen with repeating determinants such as ovalbumin substituted by several DNP groups, or most bacterial polysaccharides or red cells with repeating blood

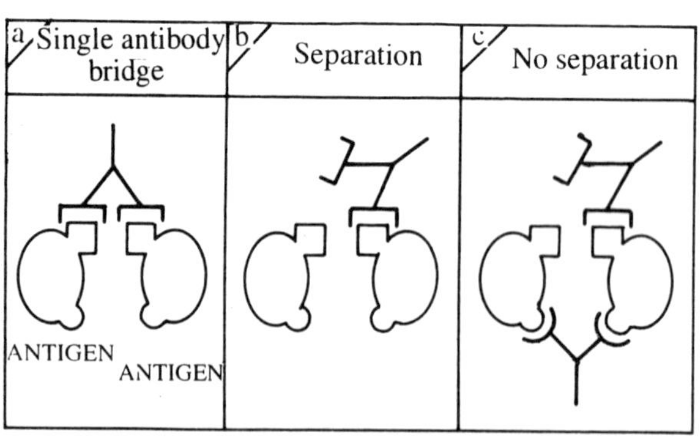

Fig. 2.5. The mechanism of Bonus Effect of multivalency

group determinants. As one moves from a univalent Fab fragment to a divalent IgG to a pentameric IgM, the bonus effect of multivalency produces striking increases in the strength of antigen-antibody complex formation.

The ability of antibodies to discriminate between different antigens was well illustrated by the range of reactivity of an anti-hapten for a series of structurally related molecules as described in Table 2.1. Since the strength of the reaction can be quantified by the affinity or avidity, the specificity of an antiserum can be related to its relative avidity for the antigens which are being discriminated. An antiserum may have a relatively greater avidity for one antigen rather than another, and this leads to the fact that the antiserum is displaying relative rather than absolute specificity; in practice we speak of cross-reactivity. An antiserum raised against a given antigen can cross react with a partially related antigen which bears one or more identical or similar determinants. In Fig. 2.6, it can be seen that an antiserum to antigen 1 (Ag_1) will react less strongly with Ag_2 which bears just one identical determinant because only certain of the antibodies in the serum can bind. Ag_3 which possesses a similar but not identical determinant will not fit as well with the antibody and the binding is even weaker. Ag_4 which has no structural similarity at all, will not react significantly with the antibody. Thus, based upon stereochemical considerations, we can see why the avidity of the antiserum for Ag_2 and Ag_3 is less than for the homologous antigen, while for the unrelated antigen Ag_4 it is negligible. It is customary to describe the antiserum as

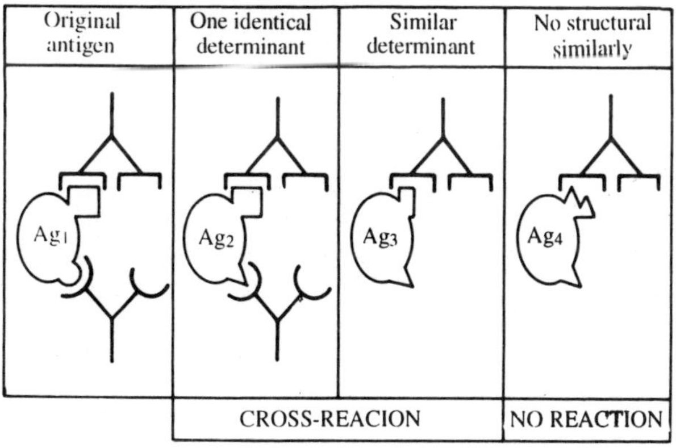

Fig. 2.6. Specificity and cross reaction. The avidity of the serum antibodies (H and H) for $Ag_1 > Ag_2 > Ag_3 >> Ag_4$
Source: Essential Immunology, I.M. Roitt; ELBS, 1988

being highly specific for Ag_1 in relation to Ag_4 but cross reacting with Ag_2 and Ag_3 to different extents.

By being directed towards single epitope on the antigen, monoclonal antibodies frequently show high specificity in terms of their low cross-reactivity with other antigens.

CHAPTER 3

Immunoprecipitation, Agglutination and Complement Fixation

Introduction

Antigen-antibody reactions usually occur in two stages. The 'Primary Interaction' is the specific recognition and combination of an antigenic site with the binding site of its corresponding antibody. This has been described in the previous chapter. Then follows a 'Secondary Interaction' which activates some effect or function. This can be, for example, complement fixation, macrophage binding or simply agglutination and precipitation. It is this secondary manifestation which is usually observed in the laboratory. It must be remembered, however, that this secondary effect is only a reflection of the primary antigen-antibody reaction and does not directly correspond to it, especially with antigens having few antigenic determinants. With monovalent haptens, for example, although the primary interaction occurs, there is no precipitation or complement fixation.

Precipitation

Precipitins are antibodies that insolubilise antigen; hence, an antiserum that produces a precipitate when mixed with antigen solution contains precipitins.

The quantitative precipitin test, developed by Heidelberg and Kendall, is the basis of all quantitative studies of antigen-antibody interaction. Increasing amounts of antigen are added to a constant amount of antibody and the weight of the precipitate formed in each tube is determined. The procedure outlined below is suitable for most antisera and antigens.

Add the following μg of antigen to a series of tubes (suitable for centrifugation): 0, 10, 20, 50, 100, 150, 200, 250, 350 and 450. Add phosphate buffered saline to each tube to a final volume of 0.45 ml. Add 0.1 ml of antiserum to each tube and mix the reactants. Incubate at 37°C for one hour and then at 4°C overnight. Spin at 3000 g for 5 minutes and remove supernatant. An angle-head rotor is to be used for this as the precipitate is formed at the side of the tube. Wash the precipitate twice by centrifugation with cold phosphate buffer-saline. Redissolve the final precipitate twice by centrifugation with cold phosphate buffered saline. Redissolve the final precipitate in 0.1 M sodium hydroxide and read the extinction at 280 nm and plot a graph of optical density of the precipitate against the amount of antigen added.

In each case, the amount of antigen and antibody in the supernatant can be found out by standard immunodiagnostic techniques like single radial immunodiffusion and double diffusion, which will be discussed later. The point of equivalence is that at which all the antibody and antigen are complexed in the precipitate, and this occurs just before the point of maximum precipitation (Fig. 3.1). The amount of precipitate increases after the equivalence point because of continued incorporation of antigen into the complex. Eventually, soluble complexes are formed in antigen excess and the amount of precipitate decreases. Precipitation is not only depend-

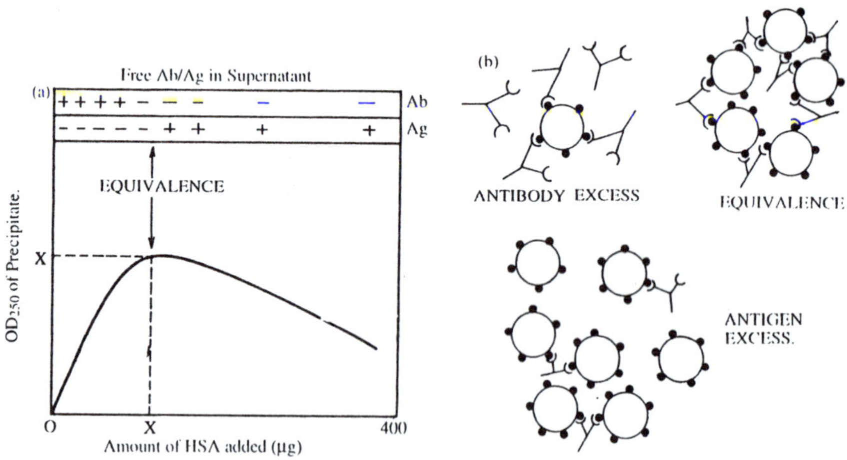

Fig. 3.1. (a) Precipitation curve of human serum albumin (HSA) with anti-HSA.
(b) Complexes formed at varying antigen-antibody ratios

ent on cross-linking, as $F(ab)_2$ fragments precipitate far less readily with antigen than whole antibody molecules.

In Fig. 3.1, at equivalence, the precipitate contains X μg of total protein. If this includes Yμg of antigen then there is (X-Y) μg of antibody. This is the total amount of antibody in the volume of serum used. Thus the quantitative precipitin test is used for quantitation of the antibody present in an antiserum.

The test can also be used for the determination of antigen valency. In the precipitin curve, when the antigen concentration is low, there is a relative antibody excess. At the other extreme, at high antigen concentrations, there is free antigen and so each combining site of the antibody is occupied. At equivalence, however, there is much crosslinking between molecules and large complexes are formed. Every antigenic determinant is likely to be covered by a separate antibody molecule at extreme antibody excess. If we calculate the amount of antibody in the precipitate at this point, we can determine the ratio of antibody to antigen, and therefore the relative numbers of molecules of each in the precipitate.

i.e. $$\frac{\text{Weight of Antigen}}{\text{Molecular weight of Antigen}} : \frac{\text{Weight of Antibody}}{\text{Molecular weight of Antibody}}$$

In a hyperimmune serum the major proportion of antibody will be of the IgG class with a molecular weight of 150,000. The molecular weight of the antigen should be known. To obtain the best estimate of antigenic valency, the ratio of antibody to antigen in the precipitate should be plotted against the amount of antigen added, as in Fig. 3.2. If the line of

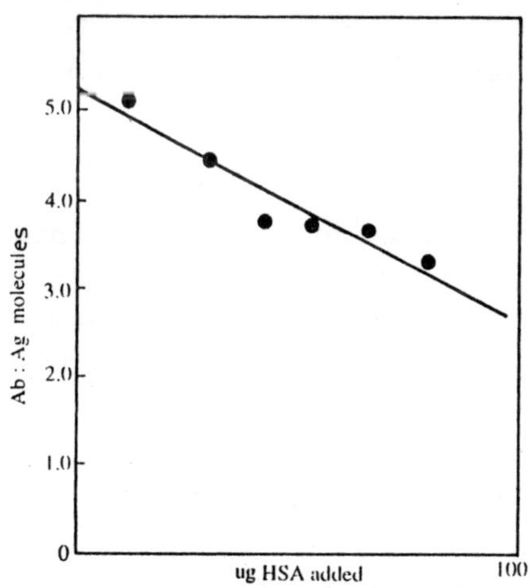

Fig. 3.2. Ratio of antibody to antigen molecules of different levels of antibody excess

graph is extended to the antibody: antigen axis, the intercept will give the ratio at infinite antibody excess. For example in an experiment using Human Serum Albumin (HSA) as antigen, the antiserum was found to recognise between 4 to 5 determinants on the antigen molecule.

The quantitative precipitin test is particularly useful as it can be carried out without antibody standards; only known concentrations of antigen are required. The shape of the curve varies with different antigens: with protein antigens a sharp peak is usually obtained, but with polysaccharides there is usually a broader peak.

In addition, different antisera have different properties. Rabbit antisera give the typical curve shown in Fig. 3.1, whereas horse antisera give precipitates which are relatively more soluble in both antigen and antibody excess (Fig. 3.3). Both types of precipitin curves may be obtained with human antisera.

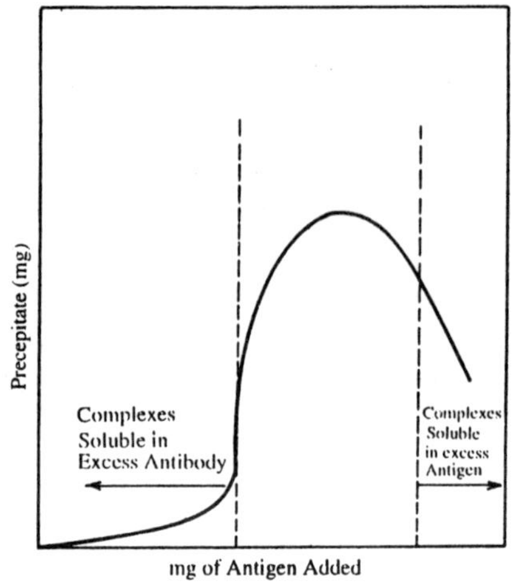

Fig. 3.3. Precipitation curves of horse antibody with antigen

Although the precipitin test gives a great deal of basic information, it is lengthy to perform and requires relatively large quantities of reagents. In addition, it is difficult to compare two antisera or two antigens qualitatively. Many gel precipitation techniques have been developed to overcome these problems.

Single Linear Immunodiffusion

Diffusion in gels has been used since 1905 and has been developed into highly sophisticated technique. One of the oldest and best characterised of immunodiffusion techniques is the classical linear single diffusion test.[8,9,10] Single diffusion tests are those in which one of the two reactants, either antigen or antiserum, remains stationary (the internal reactant) while the other (the external reactant) diffuses into and complexes with it.[11] These are performed in two ways: the single linear immunodiffusion and the single radial immunodiffusion. In linear tests, a small tube is partially filled with agar gel mixed with a small amount of antiserum as internal reactant and then, after this has gelled, the external reactant, namely the antigen solution is used to overlay it. Antigen is used in large immunologic excess relative to antibody. Hence, immediately after the antigen is layered upon the antiserum column, the tendency for antibody to diffuse upward into the antigen is suppressed, and the antigen, instead, diffuses downward into the antiserum.

Downward diffusion of the antigen quickly dilutes it at its advancing front from initially too great an excess to be precipitated by the antiserum to what later becomes a precipitating ratio with the antiserum. Consequently, a short distance below the interface of the two reactants, a precipitate begins to appear. More antigen feeds into and passes through this area, diluting itself by diffusion while encountering essentially unchanging concentrations of antibody in succeedingly lower planes of the antiserum-charged agar. When the antigen diluted by its diffusion precipitates with antibodies at the antigen diffusion front, it is being followed by continued diffusion of more antigen in higher concentration behind it. Consequently, as free antigen feeds into the back of the antigen diffusion front where unstabilised precipitation is developing, it tends to dissolve the developing precipiting band from behind. Thus, there is precipitation at the diffusion front and partial reversal of this behind the front, so that the precipitation band is, in fact, a wave of precipitation moving through the internal reactant with a sharp front and a trailing edge which is ragged in rabbit antisera (R-type) and fairly sharp in horse antisera (H-type), since in the latter, the antigen-antibody complexes both in "antigen excess" and "antibody excess" are soluble as discussed earlier.

Antiserum in a single diffusion tube may have antibodies to several antigens constituting the antigen mixture being analysed. Each antigen-antibody system will react independently of the other because the biophysical and immunologic characteristics of each system are unlikely to be identical with those of any other. Hence, two different systems usually precipitate in different planes of the reaction column and continue to do so during most of the test period. Thus, if a mixture of antigens A and B is diffused into gel containing antiserum to both, the two may first precipitate in a single broad band immediately below the antigen-antibody interface. But as time passes, they will be distinguished from each other by a progressive séparation of their two separate bands. Each of these will migrate through the antiserum gel independently of the other and at its own characteristic rate, as determined by factors such as concentration of the antigen forming it, the antigen's diffusion rate, concentration of antibodies forming it, and miscellaneous properties of the immunologic reactions between the antigen and its various antibodies. Theoretically it is very unlikely that for any two antigen-antibody systems all these factors will be exactly the same. But practically, the distinction between two different systems tends to blur in single diffusion tubes, especially when R-type antibody is being employed, because the trail of the leading wave of precipitation tends to extend back over the front of the following wave of precipitate. This is why various double diffusion or immunoelectrophoresis tests are preferred for qualitative analyses.

Although two entirely independent antigen-antibody systems will not interfere with each other, cross-reacting systems will. This may be either troublesome or helpful, depending on the test situation. It would be useful, for example, in determining whether a solution contains antigen A or antigen B. For this, the solution is mixed with each of the two antigens separately and this is compared with control charged with antigen A or antigen B alone. If antigen A is present in this solution, then its concentration in the solution will be added to that of its supplementary antigen, and the rate of migration for the band in the tube receiving antigen A will be accelerated while its migration rate in the tube receiving antigen B will remain unchanged. (Fig. 3.4).

Consider an antigen with determinants x and y and another with only determinant x. Antigen xy diffused into gel containing antiserum to determinants x and y (i.e., antiserum to antigen xy) would form one precipitin band. Antigen x diffused into this antiserum would similarly form one

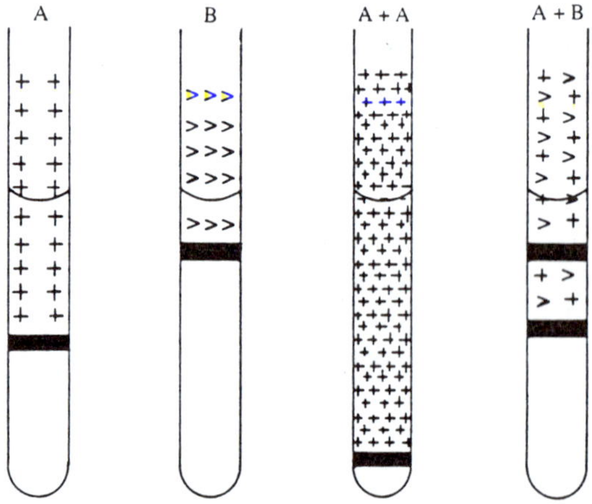

Fig. 3.4. Effects on precipitin band formation of mixing together of identical or different antigens (+, > are antigens)

precipitin band. But, suppose antigens xy and x were mixed for testing in three different proportions to each other: xy predominating, xy and x at equivalent concentrations, and x predominating. With xy predominating, only one band would form, consisting of mixed aggregates of xy and antibodies to determinants x and y. No uncomplexed antibodies to either determinants would remain to react with antigen x trailing behind. With equivalent concentrations of each, there would still be only one precipitin band formed, consisting of a mixture of antibodies to both determinants complexed with both antigens. But the apparent diffusion rates of antigen xy would be increased (as they would also for x, though proportionally less) because x, by "aiding" antigen xy in complexing with antibodies to shared determinant x, would be decreasing the number of antibodies through which antigen xy must diffuse. Two precipitin bands may be seen with x predominating—that developed by x at the front and migrating as though no antigen xy were present, and that constituted by xy trailing behind and composed of antigen xy and antibodies only to its y determinant, since all antibodies to x would have been consumed by reaction in the more advanced front of antigen x (Fig. 3.5). Only one precipitin band still might form if the concentration of antibodies to y relative to that of antibodies to x were very low in the antiserum being used. Then, even at increased relative proportion to antigen xy, antigen x still cannot advance far enough ahead of it to form a detectable independent band.

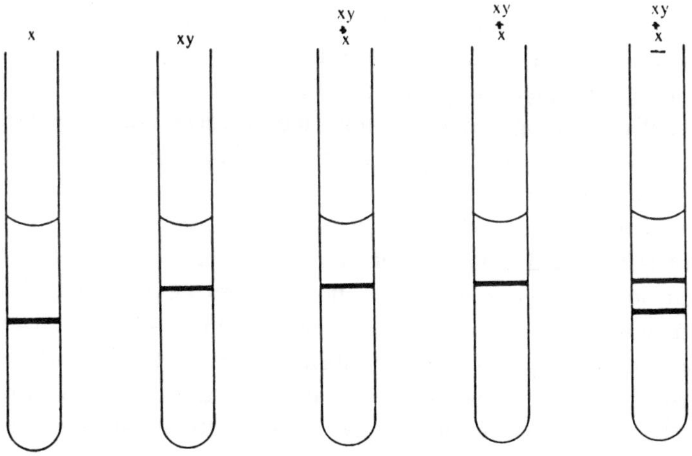

Fig. 3.5. Various possibilities for precipitin band formation in single diffusion tube tests with crossreacting antigens x and xy and antiserurm containing the corresponding antibodies, when the two antigens are used singly or mixed in different proportions (underlining for antigen indicates its predominance)

Source: Immunodiffusion, Crowle; Academic Press, 1973

From the point of view of Immunodiagnostics, the linear tests are not preferred. In fact, in most cases double diffusion tests are only used. However, one form of the single diffusion test which is mostly used is the single radial Immunodiffusion test.

Single Radial Immunodiffusion (SRID)

In a radial single diffusion test, the external reactant (e.g., antigen/antibody) in serologic excess to the immediately surrounding internal reactant is deposited in a well cut-in, or at a point on, a thin flat plane of internal reactant — containing medium. From this it diffuses radially into, and combines with, the internal reactant to form a disc of precipitate that expands out into the surrounding medium. When it attains maximum size, the front of the diffusing external reactant reaches serologic equilibrium with the surrounding internal reactant and the disc stops growing[12]. Compared to other immunodiffusion tests, radial single diffusion tests have become very popular because they offer the unique opportunity of quantitating antigens and antisera with great ease.

In this test, the final size of the disc of precipitate has a linear relationship with the initial concentration of the external reactant and is inversely related to the internal reactant concentration.[12,13] (Fig. 3.6). In the linear test, the internal reactant functions primarily as an indicator for the progress of the external reactant diffusion and is usually not expected to precipitate the external reactant to slow down its diffusion. The linear test, therefore, provides primarily physical data from which are derived qualitative and quantitative characteristics of the external reactant. By contrast, in the radial test, migration of the external reactant is inhibited significantly and progressively by the internal reactant until it is stopped, when by the dilution and attrition of its preceding migration through the internal reactant and the medium, its concentration comes to equilibrium with that of the internal reactant. The more concentrated the external reactant is, the further its disc of precipitate will expand before being arrested; and the lower the concentration, the less the precipitate disc will expand. Similarly, the weaker the internal reactant, the furthest the external reactant can migrate before equilibrium is reached. A basic difference between linear and radial tests is, that the former provides primarily physical and the latter serologic data.[14] Hence, these tests are of greater clinical interest. As a result of this difference, the radial test is more independent of physicochemical influences.

The linear relationship between the concentration of the antigen, and the size of the precipitate in a radial test will fail if the antigen population is physically heterogeneous so that it occurs in different proportions of

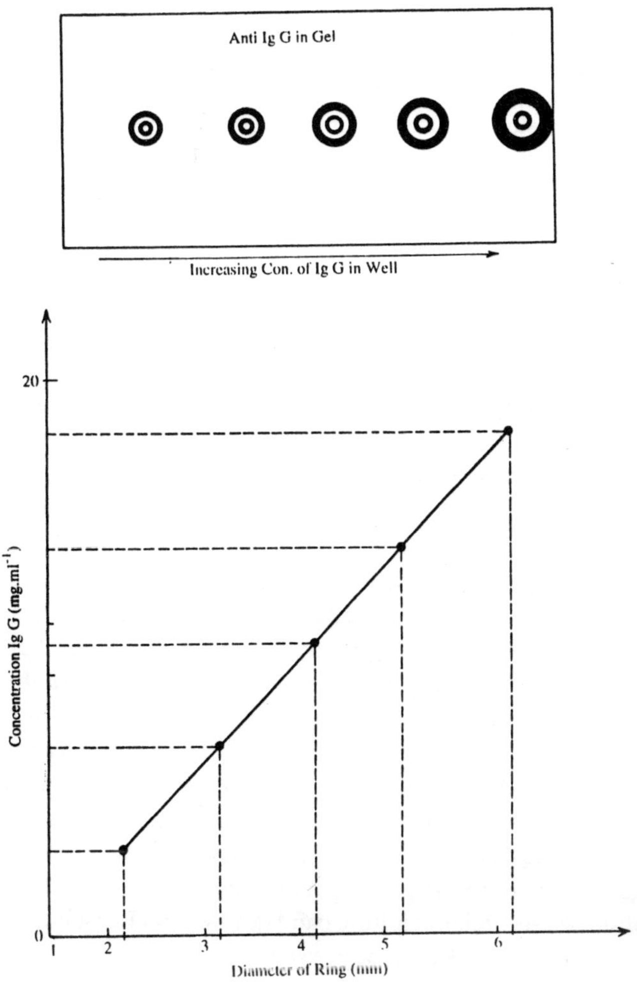

Fig. 3.6. Linear relationship of precipitin ring size and concentration of antigen in SRID for IgG

subpopulations from sample to sample; if there are concentration-dependent interactions between antigen molecules, antigen and accompanying solutes, or antigen and Immunodiffusion medium; or if the antiserum is not monospecific for the antigen but contains antibodies which also can react with contaminating antigens.[15, 16]

Since data from the radial test are serologically defined, they must be related to serologic standards; either the standard must be serologically identical to the unknown, or differences between them must be recognised and compensated for, to obtain accurate and reliable results. This statement can be understood in the light of the following observations:

1. Antiserum to IgA standardised for serum IgA will give falsely low absolute values of salivary IgA because the latter is a larger molecule.[17]

2. Antisera standardised against a predominantly K-type IgG show major differences in quantitations of IgG in general, due to possibly different proportions among these antisera of precipitins with λ specificity.[18]

3. Erroneously high values of IgM were traced to the presence in the patient's sera of low-molecular weight varieties of IgM.[19]

Although the linear relationship between Ag-Ab concentration and the area of precipitation holds over a wide range of reactant concentrations, it can fail in extremes. Also, the size of the precipitation disc should be read only after allowing enough time for the Ag-Ab reaction. Premature reading will give erroneous results. The best results are obtained in quantitative radial tests when monospecific antiserum is used against a homogeneous antigen, and hence only one precipitating system is present. When antigen and antibody precipitate each other, they create a lack of their respective molecules on either side of the immediate area of precipitation known as a "Sink". If antigen is in moderate instead of slight excess, the antigen sink will be absent whereas the antibody sink will be more prominent than it is in balanced conditions, because antibody molecules are not diffusing into the reaction area fast enough to replace those precipitated by excessive antigen. Under these conditions, free antigen will diffuse not only through the forming precipitate but also some distance beyond it before encountering enough antibody to precipitate, resulting in the formation of a second precipiting band (known as a Lisegang precipitation) by this single antigen-antibody system.

Thus, in single radial immunodiffusion, when multiple rings of precipitate develop at constant temperature in a test involving a single precipitating system, they may be the periodic precipitates of Lisegang and this is explained as the "sink effect".

The SRID can be used in comparative tests in which antigens and antisera are compared directly, although for practical purposes it is the double immunodiffusion which is used. When identical antigens are deposited fairly close to each other in an antiserum charged gel, they will influence the growth of each other's discs of precipitation. If they are barely within each other's spheres of influence, the disc of precipitate for each bulge toward the other; if they are closer together, their discs will be flattened across the plane between them. This is because in single diffusion tests, the internal reactant is not as static as the name of these tests implies,

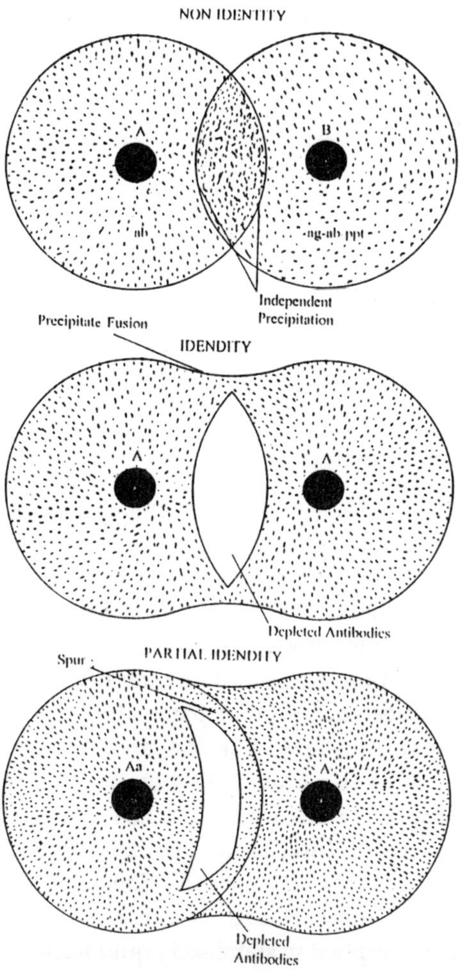

Fig. 3.7. Reactions of nonidentity, identity and partial identity in SRID comparitive tests (A, Aa, B are antigens)

for it diffuses toward the advancing front of antigen (called "back-diffusion"). The various patterns obtained when identical, partically identical and nonidentical antigens are used in comparative tests as shown in Fig. 3.7 and can be explained as follows:

1. In case of a reaction of non-identity, the precipitations are independent, that is, there are two independent reacting systems and so the zones of precipitation overlap.

2. In a reaction of identity, the precipitation discs merge but do not overlap and this is because of the back-diffusion which makes both the precipitin bands advance in opposite directions, i.e.,

towards each other. The fusion zone of precipitate, however, does not have any precipitate since the sink effect depletes the antibodies in this region.

3. In a reaction of partial identity, a region of depleted antibodies is seen but two different precipitates are seen independent of each other and there is no fusion of the zones of precipitation but one zone develops into the other which extends over it, as shown in the figure. Since there is partial identity, spur formation is seen around the region of depleted antibodies and beyond this region, a secondary precipitation is seen which is due to the nonidentical fraction of the antigen preparation.

Immuno Quanton Plate System

One of the most important diagnostic applications of single radial immunodiffusion is the quantitation of class specific immunoglobulins in human serum. The *'Immuno quanton Plate System'*, a product of Span Diagnostics, Udhna, India, is a readymade SRID diagnostic kit for this purpose. In this kit, immunoquanton plates are provided which are of buffered agarose medium containing the internal reactant (specific antibody). There are three such type of plates: IgM, IgA and IgG with the internal reactants as Anti-IgM, Anti-IgA and Anti-IgG, respectively. A reference serum vial is also provided whose levels of IgG, IgM, IgA are mentioned along with confidence ranges. IgA of 206 mg% concentration (confidence range 175-237 mg%), IgG of 2012 mg% (confidence range 1711-2313 mg%) and IgM of 190 mg% (confidence range 162.5-218.5 mg%) are contained in the reference serum and it is prepared from healthy adult donors and is supplied in stabilised liquid form.

The test serum (sample) is diluted appropriately (1:2 or undiluted for IgM, 1:4 for IgA, and 1:8 for IgG determination) and quickly and carefully added to the wells in the plate using a 5 micro litre syringe, after removing the lid of the plate. Then, the lid is replaced and the plate incubated at 25-30°C in a humid chamber, which can be easily made by placing moist filter paper or a piece of moist cotton at the bottom of a plastic box or a petridish. The plate for IgG estimation is incubated for 48 hours, while for IgM, IgA estimation it is incubated for 72 hours. After incubation, the diameter of each precipitin ring is measured accurately (upto 0.1 mm accuracy) with the help of the Quantometer, a special measuring scale. The class specific immunoglobulin concentration corresponding to the measured diameter are read directly from the reference value chart provided for the particular batch of immunoquanton plates. Such results are accurate and acceptable only when the value obtained for the reference serum is within the confidence range of the reference serum. If the value so obtained for the reference serum does not fall within the confidence range, it is necessary to prepare a calibration curve to obtain the true assay value. Immunoquanton plates of IgA, IgG and IgM are stable at 2-8°C and are not to be frozen. They use a heavy chain specific antiserum to the specific immunoglobulin. Sodium azide (0.1%) and thiomersal (0.01%) are added as preservatives. Serum samples showing marked bacterial contamination are not to be used for the test.

Immunoquanton plate system is a simple and reliable system for quantitating immunoglobulins for the diagnosis of monoclonal gammopathies and other related abnormatities. Determination of immunoglobulin levels at different times during the course of a disease is important for monitoring the therapeutic response in patients with monoclonal gammopathies, certain liver diseases and several other diseases with altered gammaglobulin levels. For example, this is the system which is used in most of the Indian laboratories for the heavy chain typing of multiple myeloma and for prognosis of the disease. The clinical significance of different immunoglobulins is given in Table 3.1.[20,21,22]

Table 3.1 Clinical significance of immunoquanton plate system in the diagnosis of certain diseases

Type of Immunoglobulin	Normal serum concentration of ab (mg%)		Increased in	Decreased in
	Adults	Children		
IgA	60-490 (mean: 210)	50-100	1. IgA Myeloma 2. Cirrhosis 3. Chronic infections	Myelomas of other Immunoglobulin classes, Protein losing conditions and agamma globulinaemia
IgG	700-1800 (mean:1200)	350-800	1. IgG Myeloma Monoclonal 2. Chronic infections polyclonal 3. Liver diseases-polyclonal	Myelomas of other Immunoglobulin classes, Protein losing conditions and agamma globulinaemia
IgM	60-280	15-140	1. Waldenstrom's macro-globulinaemia 2. Chronic infections 3. Liver diseases	Myelomas of other Immunoglobulin classes, Protein losing conditions and agamma globulinaemia

Double Diffusion Test

Immunodiffusion tests in which both antigen and antiserum diffuse toward each other significantly are known as double diffusion tests. Like single diffusion tests, these tests can be radial or linear. In addition, because both reactants diffuse, they can also be angular.

As the antigen and antibody migrate towards each other in a gel, a line of precipitation is formed where the two reactants meet. As this precipitate is soluble in antigen excess, a sharp line is produced at equivalence, its relative position being determined by the concentration of antigen and antibody in the agar. The local concentration of each reactant depends on:

(a) Absolute concentration in the well

(b) Molecular size of the reactant

(c) Diffusivity of the reactant through the gel

If the antigen and antibody contain several molecular species, multiple lines of precipitation will be formed.

In actual experiment, 2% agar is made in barbitone buffer and allowed to set on a glass slide to give a layer of uniform thickness. Using a gel punch, a pattern of wells is punched out in the agar to hold the reactants. The reactants are placed in the wells and the precipitation reaction allowed at a constant temperature in a humid chamber overnight. The slides are then soaked in 1% tannic acid for 1 minute to increase resolution or may be washed in PBS for 24 hours and stained using a protein stain such as *Coomassie brilliant blue or amido black* and then destained to see the lines of precipitation. The latter is a permanent preparation.

There are three basic patterns of precipitation as shown in Fig. 3.8.

1. *The Reaction of Identity* occurs between identical antigenic determinants, the lines of precipitation fuse to give one continuous arc.
2. *Reaction of Non-Identity* occurs where two antigens do not contain any common antigenic determinants, the two lines are formed independentaly and cross without any interaction.
3. *Reaction of Partial Identity*, which has two components:
   (a) Those antigenic determinants which are common to both antigens give, a continuous line of identity,
   (b) The extra determinants on one of the antigens gives, in addition, a line of non-identity, so that a spur is formed. Of course, there may be unique determinants in both antigens, this would give rise to two spurs.

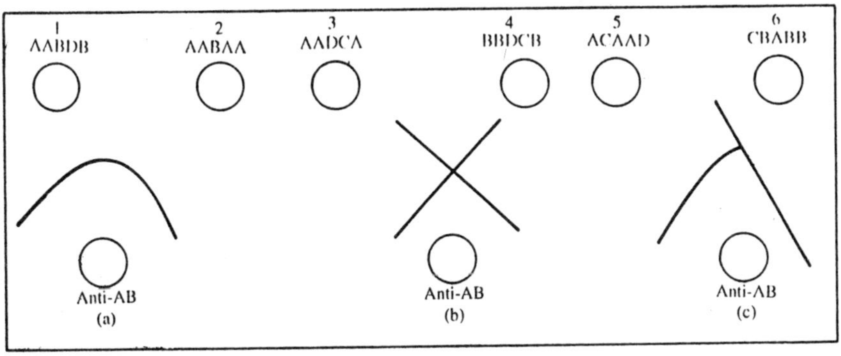

Fig. 3.8. Explanation of basic patterns of precipitation obtainable in Ouchterlony Immunodiffusion, A-D represent antigenic determinants present within a series of protein molecules. 1-6, used as antigens. An antiserum has been raised against two of these determinants. (Adapted from *'Practical Immunology'* by Hudson. L. and F.C. Hay)

(a) Reaction of identity
(b) Reaction of non-identity
(c) Reaction of partial identity

In order to demonstrate the reactions of identity and non-identity, an antiserum that recognises a wide variety of antigenic determinants must be used.

The design of the *Ouchterlony technique* may be varied to obtain semi-quantitative data. Two main applications of this kind include:

(a) Determination of relative antigen concentrations.

(b) Estimation of molecular weight and diffusion co-efficient.

As the position of the precipitation line is determined by the relative concentrations of the antigen and antibody, a semi-quantitative estimate or relative antigen concentration can be obtained by comparing the position of the lines using several concentrations of antigen and a constant antiserum as shown in Fig. 3.9a.

Another property of the precipitation lines obtained in the Ouchterlony test is that an antigen with molecular weight and diffusion coefficient close to IgG tends to give a straight line of precipitation with antibody. If the antigen has a lower molecular weight and higher diffusion coefficient than IgG, the arc of precipitation curves towards the antiserum well, while if its molecular weight is higher and diffusion coefficient lower than IgG, then the precipitin line curves towards the antigen well. Fig. 3.9b shows such an effect. Here, albumin, IgG and *Keyhole Limpet Hamocyanin* have been taken as examples.

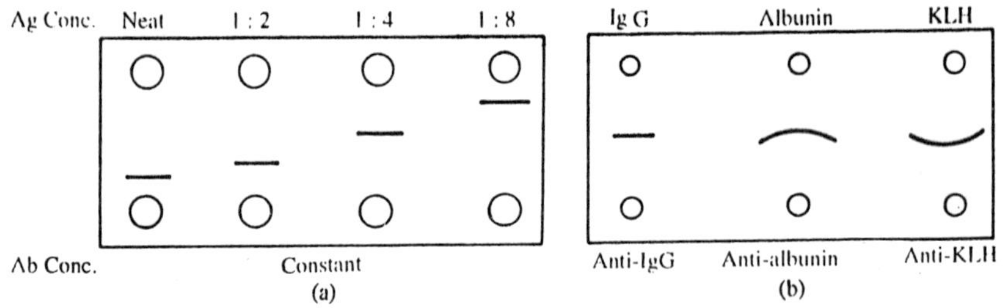

Fig. 3.9. (a) Relative antigen concentration determined by double diffusion in agar
(b) Estimation of molecular weight and diffusion coefficient by double diffusion in agar
(Adapted from *Practical Immunology*, Hudson. L. & F.C. Hay.)

Immunoelectrophoresis

The antigen-antibody interactions we have studied so far are solely by simple diffusion. Increased resolution, and often quicker results can be obtained by a combination of electrophoresis and immunodiffusion in gels.

In a typical form of immunoelectrophoresis, a microscope slide is flooded with an even layer of agar gel in an appropriate buffer (say, barbitone buffer), and the ends of the gelled agar on the slide are connected with buffer vessels. A hole is punched in the agar and charged with the antigen and, after electrophoresis, slots are cut parallel to the path of electrophoretic migration on either side of it far away to permit development of precipitin arcs in the intervening agar. An antigen fraction usually has one of three shapes (Fig. 3.10): it is round or symmetrically ellipsoid, it is drawn out on either side of its centre equally or unequally, or it assumes the shape of a comet with a trail either before or behind it.[24] Its shape controls the shape of the precipitin arc formed with antibody diffusing against it. After antibody is placed in the trough, it diffuses towards the path of electrophoresis as a straight front. Meanwhile, each fraction which has been electrophoresed expands radially by difusion from its source, part of it moving toward the antibody as a curved front corresponding to the original curvature of the spot of electrophoresed antigen. Thus, if the final shape of the antigen spot is round, it presents a semicircular front; if it has the form of a symmetrical ellipse, its front will be symetrically ellipsoid; and an antigen spot that is asymmetrical will present a corresponding asymmetrical curved front. The acuteness of curvature also

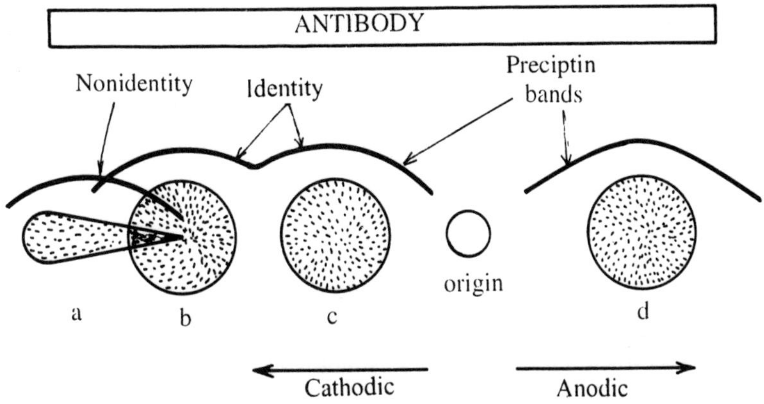

Fig. 3.10. Precipitin band patterns in a typical immunoelectrophoresis experiment. Reaction of identity (gull-wing formation) and nonidentity are shown a, b, c and d are antigen components in the mixture used. (Adaped from *Immunodiffusion* Alfred J. Crowle, Academic Press)

depends largely on the relative diffusion rates of antigen and antibody (in turn depending on their respective concentrations and molecular weights), for if the antigen does not diffuse very rapidly from its origin by comparison with the rate of antibody diffusion, the precipitin arc formed by it will curve around it more acutely than if opposite conditions prevail.

Some antigens produce a single precipitin band only by double diffusion tests but on immunoelectrophoresis divide into two distinctly different fractions. If these have remained close enough during electrophoresis, each forms its own arc, but the tips of these arcs fuse to form the so-called "gull-wing" pattern.[25]

The form and location of precipitin arcs in immunoelectropherograms are determined by:

(a) Electrophoretic migration of the reactants

(b) Their subsequent diffusion and immunologic interaction

In actual experiment, the electrophoresis of the antigen is done at a current of 7-8 mA per slide and of 5-7 volts per cm for 60 minutes. The antibody is then filled in the trough and the slide incubated in a humid chamber overnight to allow immunodiffusion and formation of precipitin arcs. Subsequent steps of washing, staining and destaining are similar to those mentioned in immunodiffusion.

An important clinical application of immunoelectrophoresis is in the diagnosis of multiple myeloma. The serum of the patient is subjected to immunoelectrophoresis with a control of normal human serum using anti-human immunoglobulin. The two antigens (patient's serum and normal human serum) are placed in wells on either side of a trough for the antibody. The serum components are separated by electrophoresis and form precipitin bands by reaction with their corresponding antibodies. Whichever antibody is monoclonally expressed in high amounts in the myelomic serum will, because of its increased concentration, form a "bump" precipitin that can be identified as different from the arc in the normal human serum. Thus detection, as well as typing of multiple myeloma is possible. Figure 3.11 shows the immunoelectrophoresis of a serum of a IgG myelomic patient. The bump is seen in the IgG precipitation arc towards the antiserum well. A rabbit anti-human immunoglobulin was used as antibody.

In a completed immunoelectropherogram, the centre from which an antigen has diffused to form its arc of precipitate can be determined accurately by noting where a line drawn from the apex of its arc perpendicular to the path of electrophoresis crosses this path. (Fig. 3.10). The position of one antigen therefore is easily compared with that of another as a standard, both to identify the former and to determine its immunoelectrophoretic mobility. For example, to determine the mobility of an unknown antigen x, one relates it within an immunoelectropherogram to those of two others a and b, whose mobilities are known:

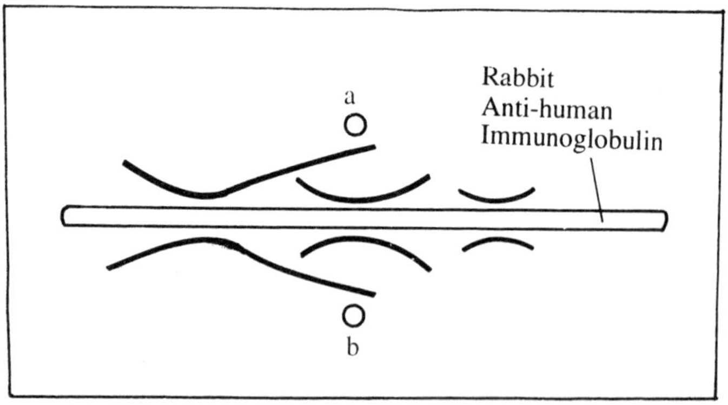

Fig. 3.11. Immunoelectrophoretic diagnosis and classing of multiple myeloma.
(a) Top well with normal human serum,
(b) Bottom well with IgG myelomic serum
Note the "bump" in the IgG precipitin band in the myelomic case

$$m^x = m^b + (m^a - m^b)\frac{x - b}{a - b}$$

in which m^x, m^a, and m^b are mobilities of the unknown antigen x, antigen a and antigen b respectively in 10^{-5} cm^2 $volt^{-1}$, x-b is the distance measured on the immunoelectropherogram between the unknown antigen and reference antigen b, and a-b is the distance measured between the two reference antigens themselves.[26] The reference antigens (markers used include albumin, IgG, ceruloplasmin, and transferrin. Albumin and IgG are usually identified by their location, intensity and shape; ceruloplasmin and transferrin can be more definitely identified by differential stains for copper and iron, respectively.[27]

Reversed Immunoelectrophoresis

Usually in immunoelectrophoresis, antigens are electrophoresed and then detected with antiserum. But in reversed immunoelectrophoresis the antibodies of an antiserum are electrophoresed and then are detected with antigen. The goal here is to identify electrophoretically the precipitating types of immunoglobulin.[28,29] The reversed immunoelectrophoresis is more difficult than conventional immunoelectrophoresis. Its resolution is poorer because most precipitins are closely related to γ_1 and γ_2 globulins, and because individual types of precipitins tend to be electrophoretically heterogeneous.

Though agar and agarose are the media commonly employed for immunoelectrophoresis, other media polyacrylamide, polyacrylamide-agarose, cellulose acetate, gelatin, starch gel and even paper have been used for the technique. Since agar and agarose gels yield excellent results and are easy to obtain and work with, they are preferred.

Immunoelectrofucussing

Another technique of immunoelectrophoresis widely used is immunoelectrofocussing. The principle of isoelectric focussing is to separate populations of charged antigens from each other by rapidly creating a stable pH gradient in the electropherogram so that each population will migrate to an area corresponding to its isoelectric point. Isoelectric focussing can be used instead of zone electrophoresis to separate antigens before they are precipitated by their antibodies. This is known as, immunoelectrofocussing. It can be done in two ways:

(a) Two-step procedure

(b) One-step procedure

In the two-step procedure, antigens are electrofocussed in a column of polyacrylamide, and then this is embedded in agar for immunoprecipitation with antiserum.[30] In the one-step procedure, a microscope slide is covered with 1.5% agarose gel prepared in 2% ampholine (trademark name for mixtures of small peptides which create the required pH gradient during electrophoresis), at pH 3 to 10, charged with human serum, and electrophoresed for 5 hours at 15 volts between a 2% ethylenediamine cathode buffer and 1% posphoric acid anode buffer. After electrophoresis, the slide is submerged in neutral 1M phosphate buffer for 4 minutes, and adjusted throughout to the same pH, and finally drained. Then, antiserum wells are filled, and precipitin arcs allowed to develop as in conventional immunoelectrophoresis.[31]

Cross-over Electrophoresis

Cross-over electrophoresis is another immunoelectrophoretic technique based on the principle that γ globulins are exceptional in their cathodic migration while most other proteins move to the anode. This property is used in cross-over electrophoresis to cause antibody and antigen to migrate towards each other in gel and form lines of precipitation. The technique is similar to a simple Ouchterlony test but is much faster because it is electrically driven, and more sensitive as all the antigen and antibody are moved towards each other rather than diffusing radially. The precedure involves the punching of two wells in an agar gel slide. Antibody and antigen are placed in the anodic and cathodic wells (Fig. 3.12) respectively, and the slide is run in an electrophoresis tank for 15 minutes and examined. The technique is used in rapid clinical screening procedures, for example, the detection of Australia Antigen.

Rocket Immunoelectrophoresis

Another important immunoelectrophoretic technique is 'Electrophoresis in Antibody-containing media' which is also known as 'Rocket Immunoelectrophoresis'. Just as cross-over electrophoresis is related to Ouchterlony double diffusion, electrophoresis in antibody containing media is related to the Single Radial Immunodiffusion (SRID) test. As in the SRID test, the antiserum is incorporated into agar and wells are cut to hold the antigen. When an electric current is applied, the antigen migrates anodally into the agar while the antibody migrates cathodally. At first, soluble complexes are formed in antigen excess. As the antigen migrates further, it becomes more dilute because antigen is held back in complexes, eventually equivalence is reached and a precipitate is formed. The precipitate redissolves and moves forward as more antigen reaches

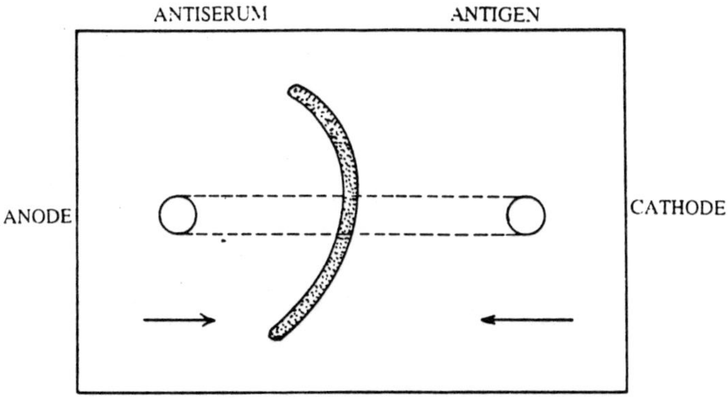

Fig. 3.12. Cross-over electrophoresis

it. Finally when no more antigen remains to enter the precipitate, a stationary stable arc is formed. Rocket shapes of precipitation are usually formed (Fig. 3.13) and the area under the rockets is proportional to the concentration of antigen.

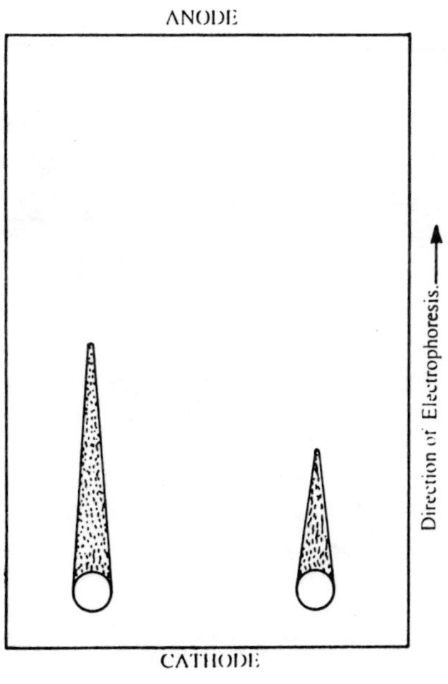

Fig. 3.13. Rocket electrophoresis in antibody containing media

Two-Dimensional electrophoresis

Immunoelectrophoresis can also be done in two dimensions. In two-dimensional immunoelectrophoresis, the mixture of antigens to be analysed is electrophoresed on a square slide of agar or agarose first in one direction and then in a second direction at right angles to the first. If the electropherogram is fixed and stained for proteins, it will show an electrophoretic separation of antigens similar to that following electrophoresis in just one dimension, except that this pattern will extend across the plate diagonally. After this electrophoretic separation in the second dimension, immunoprecipitation is done. Immunoelectrophoresis in two dimensions can be done in two ways:

1. This is also known as "two-dimensional single diffusion immunolectrophoresis". After electrophoresis in two dimensions, a filter paper soaked in antiserum is overlaid on the gel. The separated antigens are precipitatted as discs (Fig. 3.14), as in the SRID test.

2. The other technique is also known as "two-dimensional double diffusion immunoelectrophoresis." Antigen mixture is placed in an origin near the centre of a square plate of gel and is

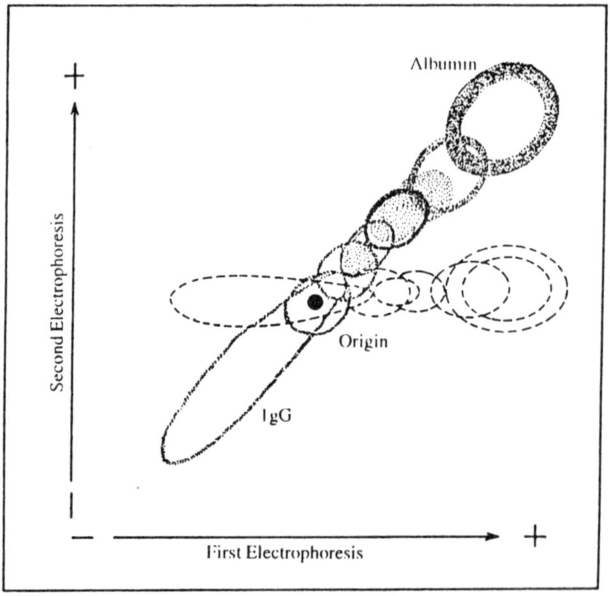

Fig. 3.14. Two-dimensional single diffusion immunoelectrophoresis

electrophoresed first horizontally and then vertically. Then four strips of fitter paper charged with antiserum are placed along the edges of the plate so that the various outward-diffusing antigens will be precipitated by corresponding inward diffusing antibodies (Fig. 3.15). Thus, in this technique, antigens and antibodies begin diffusing toward each other at 45°, instead of face to face. A major application of this technique is the simultaneous multiantigenic analysis of the precipitins in two polyvalent antisera, the top and bottom strips of filter paper being impregnated with one antiserum and the side strips with the other.

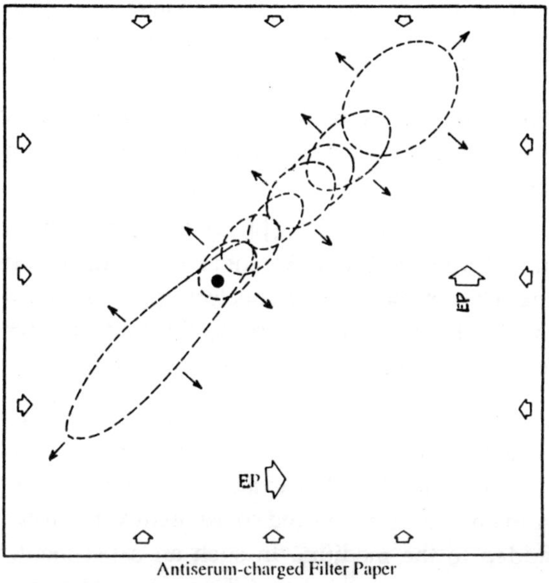

Fig. 3.15. Two-dimensional double diffusion immunoelectrophoresis
Source: Immunodiffusion Crowle: Academic Press

The information obtainable from two dimensional immuno-electrophoresis can be increased by performing primary and secondary electrophoresis under different conditions. Thus antigens of different molecular size but of similar electrophoretic mobility can be separated from each other during the second electrophoresis by using a higher concentration of agar or a sieving medium like polyacrylamide.

Enzymo-immunodiffusion

Any readily recognisable property of a constituent of antigen-antibody complexes can serve as an auxiliary indicator or these complexes in immunodiffusion tests. Enzyme activity is one such property. This characteristic is used most directly when the antigen itself is an enzyme, and antigen antibody complexes thus exhibit anzymatic activity. The utility of enzymes as indicators of antigen-antibody complexing is illustrated when they are used to detect invisible or nonstainable antigen-antibody aggregates. The following are some of the applications of enzymo-immunodiffusion:

1. An invisible aggregate of human red blood cell lysate carbonic anhydrase with its antibodies could not be detected by a conventional protein stain but was made visible by the anhydrase converting a cobalt indicator to black cobalt sulfide.[32]
2. A trace precipitate of cereal amylase (α-amylase) and its antibodies too insignificant to be stained was detected by starch hydrolysis reaction.[33]
3. An invisible, unstainable aggregate of chicken serum esterase was revealed by the esterase acting upon an appropriate substrate.[35]
4. Lactate dehydrogenase (LDH) was detected by enzymo-immunodiffusion. Five different arcs of antigen-antibody reaction could be detected in immunoelectropherograms of rat bone marrow by the effects of their respective enzymes, when there were no corresponding visible arcs of precipitate.[35]

Another form of enzymo-immunodiffusion is by studying enzyme antigens by the physical or biochemical inhibition that can be exerted against them by antibodies. For example, by precipitating an enzyme, antibodies can limit its diffusion and consequently its effect on appropriate substrate added to the medium. In such an experiment, egg yolk, for example, was used in agar to detect lipase, and its corresponding antiserum.

Special Double Diffusion Tests

There are many varieties of double diffusion tests, each being developed for a specific application. A few such tests are described below.

1. Double Diffusion after Electrophoretic Application of Antigen

Reactants usually are applied to the gel mechanically; that is, a well is charged, or a reactant is deposited on the surface of the gel. But they can also be applied electrophoretically for such purposes as separating the reactant from substances in solution with it may interfere with its precipitation. If double diffusion is to be used to detect serum antigens in a hemolyzed sample, this technique is applied. The aim here is to separate the hemoglobin from the other serum components because the former may nonspecifically precipitate in the agar.[36] For this, the hemolyzed serum is electrophoresed, just to separate the hemoglobin from the other serum components and then allowed to diffuse and react with the antiserum. The technique is different from immunoelectrophoresis since the attempt here is only to separate the hemoglobin from the rest of the antigens in serum and not to electrophoretically separate the serum antigens themselves.

2. *Double Diffusion Employing Solutions "Prepared" during the Test*

Usually, the reactant solutions are prepared and the wells are charged with these prepared solutions in the conventional double diffusion test, but the solution can be "prepared" during the test itself. Such tests are sometimes referred to as *in vivo* tests, as evident from the following examples:

(a) The antigen well can be charged with a thick suspension of bacteria from which a soluble antigen will leach and then diffuse. Meningococci are readily typed in this way.[37]

(b) Toxigenic bacteria can be grown in immunodiffusion gel enriched with appropriate nutrients, used aseptically, and to which no antibacterial preservatives have been added. Species-specific exotoxins of the bacteria can be identified by placing an antiserum-soaked strip of filter paper on nutrient agar on which the bacteria are being grown. The test known as Elek's test has been used for detecting species-specific exotoxins of pathogenic strains of *Pseudomonas*[38], diphtheria bacillus, etc. (Fig. 3.16). In such tests, however, the medium used for immunodiffusion should not only allow bacterial growth but also permit production of the exotoxin. Peptone concentration, for example, affects exotoxin production.

(c) Miniaturised Elek's test: An antiserum-impregnated strip of sterile filter paper is laid aspectically along the centre of a sterile microscope slide, and this is overlaid with a thin layer of appropriate culture medium agar. After solidification of the agar, points on the gel can be inoculated with the bacterial samples whose toxigenicity is to be tested. After incubation at the appropriate temperature for bacterial growth, the precipitation is observed. The bacteria grow and produce exotoxins that diffuse into the gel. The antibodies, diffuse upwards from the filter paper and a precipitate forms which can be stained, dried and observed as in the conventional immunodiffusion test.[39]

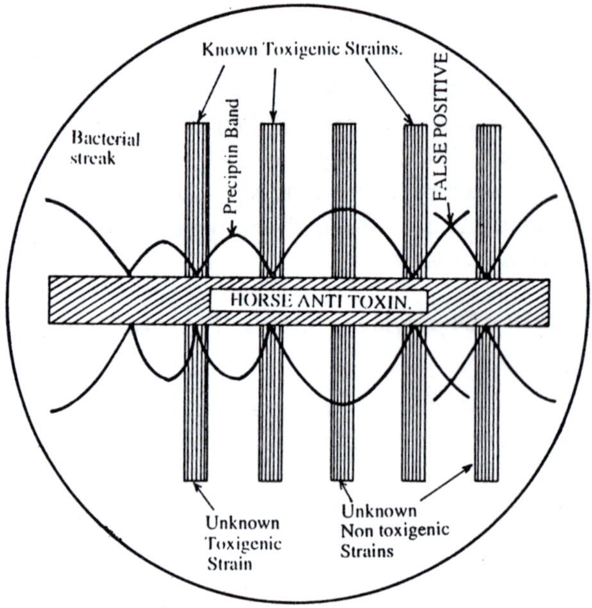

Fig. 3.16. Diagram showing how the diphtheria bacillus toxicity test is set up (see text for explanation)

(d) Intact mammalian cells and viruses can also be used in double diffusion tests. Lymph node cells from immunized mice can be cultivated in the immunodiffusion medium to detect the *in vitro* production of antibodies. In order to know if a sample has a certain virus or not, a double diffusion test can be done as follows. Two wells are punched in an appropriate medium and one of the wells is inoculated with mammalian cells and culture medium and allowed to stand for two days. The sample is then added to the mammalian cell culture established in this well and 4 to 9 days allowed for virus multiplication to proceed in the mammalian cells, and for the virus to release its antigen. The near-by well is then charged with antiserum to the virus and precipitation observed.[40]

3. *Precipitation-Inhibition Technique*

This is an indirect technique for quantitation of antigen or antibody and uses a known reference antigen-antiserum system. The principle of the technique is that the presence of antigen or antibodies is signalled by the inhibition of precipitation in a known reference antigen-antiserum system made by adding unknown to the opposing known reactant (e.g., adding unknown antigen to the reference antiserum will cause inhibition of precipitation in the reference antigen-antiserum system). The sensitivity of this test is maximized by employing only enough of each reference reactant to barely precipitate. Tumour-specific antigen could be revealed in various specimens of tissue, for example, by placing tumour antigen in a central well, charging satellite wells with precipitins to this antigen, and then mixing extracts of the various tissues with these different deposits of antiserum.[41] An extract found to contain this antigen could be further tested in serial dilution by the same technique to determine its relative concentration. One advantage of the precipitation-inhibition technique is that unlike the precipitation technique which can detect reactant of valency ≥ 2 only, this can be used to detect univalent reactants as well.

Another way of observing the inhibition of precipitation is by observing its influence on the shape or location of a forming precipitin band (Fig. 3.17). In this technique, a well charged with unknown antigen solution is set next to an area in gel where the growing tip of a precipitate between a precipitating concentration of the same antigen and appropriate antiserum will approach it, the presence of a small amount of this band; if no antigen is present in this well, precipitin band growth will continue laterally along its normal course. Similarly, antibody can be detected by this technique which will cause bending but in the opposite direction. This technique is used for the detection of bacterial endotoxins. The precipitation-inhibition technique can increase the sensitivity of the double diffusion test as much as 200-fold and is thus preferred for applications demanding sensitivity and quantification of results.

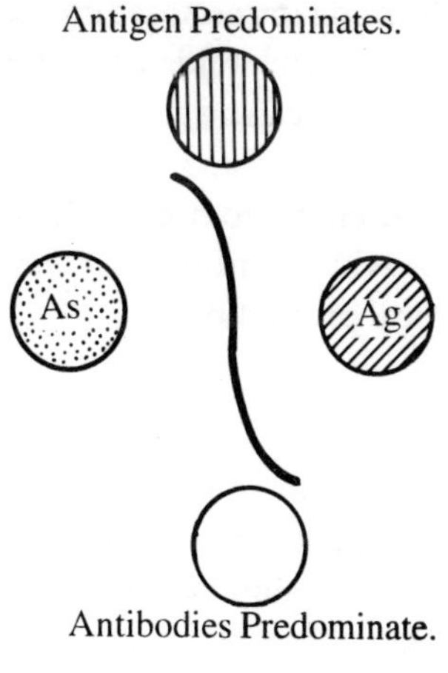

Fig. 3.17. Diagram depicting the detection of antigen and antibody by precipitin band bending

Immunosedimentation

Immunosedimentation is a two-step technique for analysing antigens in a mixture. The mixture is ultracentrifuged on a sucrose gradient, and then samples are taken from successive levels of the gradient and their antigens are detected by single diffusion immunoelectrophoresis.[42] Electrophoretic mobilty and sedimentation coefficient data can be obtained simultaneously for the varios antigens present in a mixture by this technique.

A 0.1 to 0.2 ml sample of antigen solution is layered on a 9 to 45% sucrose gradient and centrifuged at 100,000 g for 18 hours at 4°C. After centrifugation, the bottom of the centrifuge tube is pierced, and 25 successive samples are taken and transferred to a row of 25 wells cut in a line across a plate of agar gel cast as for immunoelectrophoresis, and then the electrophoresed antigens are precipitated by an overlay of antiserum. The overlay is applied by covering the electropherogram after electrophoresis with warm agar and, when this has gelled, flooding this in turn with antiserum. Precipitation is allowed to develop for 48 hours. Other techniques can also be used for applying antiserum. These include:

(a) Using an overlay of filter paper,

(b) Working the antiserum into the surface with a glass rod or a glass tube, or

(c) Applying it with a Mylar film.

Figure 3.18 shows the immunosedimentation pattern of human serum along with immunoelectrophoresis pattern for comparison.

Another way of performing immunosedimentation is ultracentrifugation in a medium containing acrylamide and photopolymerizers, so that after centrifugation the gradient can be gelled. Then, this gel can be used in a two-step immunoelectrophoretic analysis. The main advantage of immunosedimentation is that apart from providing good resolution of the separated antigens, it provides additional comprehensive information about their physicochemical properties.

Immunorheophoresis

Localized gel drying can have an accelerating effect on the development of immunodiffusion reactions. It has been observed that if the agar gel used for immunodiffusion was slightly dehydrated, precipitin band formation was much quicker. This effect can be exploited not only to accelerate immunodiffusion reactions, but also to increase their sensitivity. The principle of this technique, called "immunorheophoresis" is

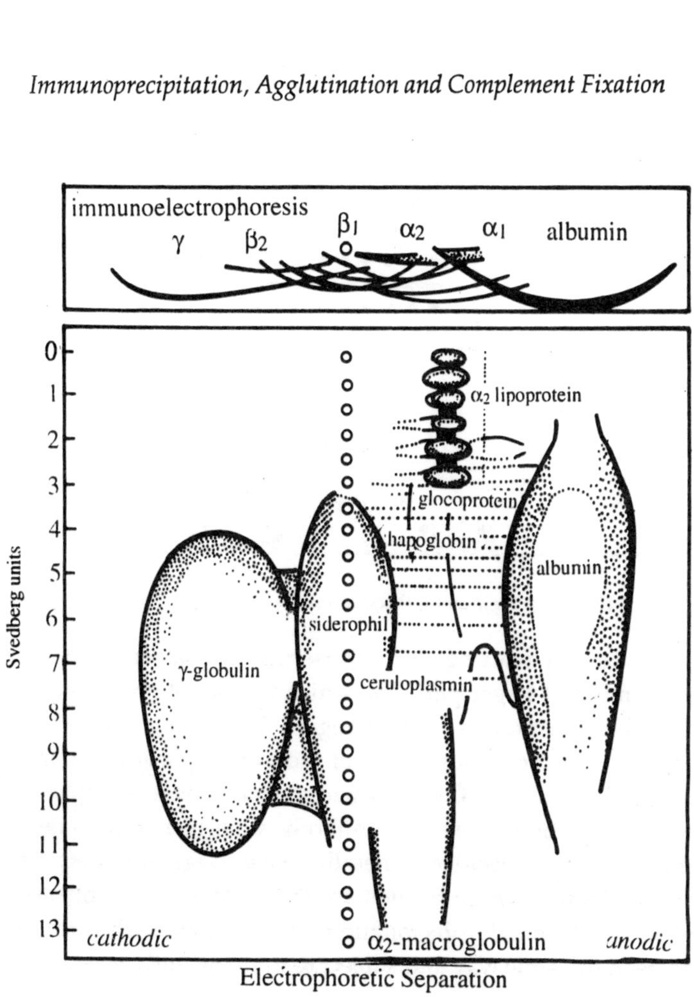

Fig. 3.18. Immunosedimentation pattern of human serum compared with Immunoelectrophoresis (above)

to perform a double diffusion plate test or immunoelectrophoresis in a vessel with a cover just slightly above the immunodiffusogram and with a slot cut in this cover above the area of gel between the two reactants[43] (Fig. 3.19). Thus, locally low humidity directly beneath this slot will cause a corresponding area of gel to dry faster than elsewhere in the immunodiffusogram. In turn, this area will attract water and, with it, reactants so that they will meet sooner and in higher concentration than in conventional immunodiffusion. Immunorheophoresis is approximately three times sensitive as compared to immunodiffusion.[43]

Agglutination

Agglutination is the same as precipitation except that the antigen is particulate rather than molecular. In liquid medium tests, the evenly

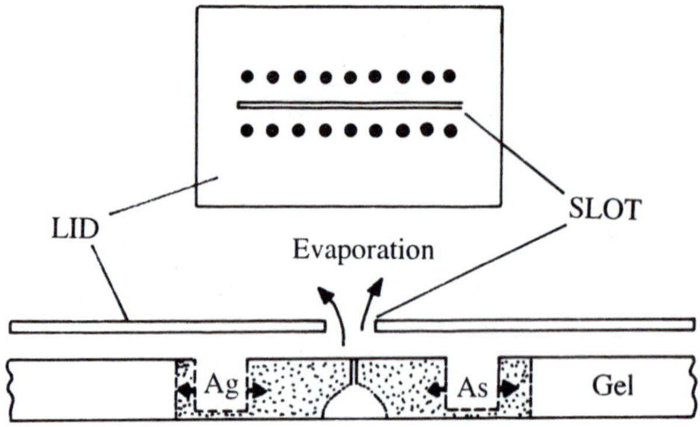

Fig. 3.19. Set up for immunorheophoresis (Ag = antigen, As = antiserum)

dispersed antigen reacting with antiserum is coarsened to become a smaller number of larger particles, which is evident both visibly and by accelerated sedimentation of the antigen. Coarsening is effected by the bridging together or agglutination of individual particles of antigen by antibody molecules into floccules whose characteristics depend mainly on the nature of the antigen. The antibody molecules have to be divalent and the antigen molecules generally have large numbers of many varieties of ligands. These ligands are the same as the antigen-receptor sites of antigen-antibody precipitation, and complex with corresponding antibodies in the same way and subject to the same influences.

The difference in antigen size between agglutination and precipitation also accounts for a major distinction between the two tests for quantity of antibody necessary to effect visible reaction. So little antibody is needed for agglutination that, practically, when antiserum is diluted to the point of inability to agglutinate antigen, it can be considered devoid of agglutinins. But an antiserum similarly diluted for precipitation still contains large numbers of precipitins, but these are simply insufficient to produce a visible precipitate. This difference in antigen size also means that in agglutination tests there is no practical range in ratios of antigen to antibody encompassing antigen excess in which agglutination is inhibited (as is easily observed in precipitation); on the other hand, inhibition of agglutination by antibody exccess is quite common (whereas inhibition in precipitation is rarely seen).

Any colloid or suspensoid with antigen ligands on it can be used as antigen in agglutination tests. These ligands may be a natural part of the particle of antigen (e.g., structural antigens on viruses, bacteria, erythrocytes; micelles of tissue lipids suspended with surfactants), or they may be placed there artificially (e.g., protein antigen adsorbed on erythrocytes). When they are natural, agglutination is said to be active; when they are artificial, agglutination is said to be passive. Apart from erythrocytes, other inert materials like latex particles or kaolin can be used for adsorbing protein antigens in passive agglutination.

Haemagglutination is the agglutination of erythrocytes by anti-erythrocyte sera. Erythrocytes may be complexed to soluble antigens by various methods and agglutinated by antisera to the coupled antigens. Two techniques are most commonly used for coupling of antigens to erythrocytes:

1. Spontaneous uptake

Erythrocytes will adsorb polysaccharides to their surface during incubation. Although this is a noncovalent binding, there is very little leaching of the antigen during the test.

2. Coupling to chemically modified erythrocytes

Erythrocytes can be chemically modified by a variety of reagents. The most common technique of coupling used is "Tanned Erythrocytes", which utilise tannic acid. Sheep RBCs are generally used. The SRBCs are washed with 40 volumes of saline by centrifugation (300 g for 10 minutes) and a suspension of the RBCs made to give a final concentration of 4% (v/v) in phosphate buffered saline at pH 7.5. 2.5 mg tannic acid is added to 50 ml of PBS* and mixed with the 4% SRBC suspension. The mixture is incubated at 37°C for 15 minutes and the cells are spinned very gently (100 g for 20 minutes). If the cells are pelleted too quickly, they will agglutinate. The cells are divided into two portions and each is washed with 50 ml PBS (100 g for 20 minutes). One aliquot will be used for antigen coating and the other as control cells. One aliquot is resuspended in 50 ml PBS, and 50 ml of a 2 mg/ml solution of protein antigen is added. This is followed by incubation at 37°C for 30 minutes. The cells are then washed in PBS by gentle centrifugation and resuspended in 100 ml of borate-succinate buffer. The second aliquot of cells is resuspended in 100 ml of borate-succinate buffer. The control cells are not coated and are used both to absorb the test antisera and as control cells in the assay.

10 ml of 40% formalin is then added to both cell suspensions while stirring. The formalin must be added dropwise during 20-30 minutes. They are then left overnight at 4°C and a further 10 ml of formalin is added to the suspensions. The cells are then left to settle for 24 hours and the super-

* PBS = Phosphate buffered saline

natant is poured off. Then a large volume of borate-succinate buffer is added to the suspensions. The cells are then left to settle for 24 hours and the supernatant is poured off. Then a large volume of borate-succinate buffer is added and the cells are resuspended by vigorous shaking. The borate buffer wash is repeated again after allowing the cells to settle for another 24 hours. Both the cell suspensions are adjusted to 1% v/v and 0.2% formalin is added as preservative. The cells can be stored at 4°C for upto two years. This protocol explains the chemical modification of erythrocytes by tannic acid and subsequent coupling. It can be used for most protein antigens. Antigens may also be coupled to erythrocytes using bisdiazotized benzidine, glutaraldehyde and chromic chloride.

Haemagglutination test is performed using a microhaemagglutination tray with 96 wells (96 well microtitre plates). One drop (0.025 ml) of PBS is added to all the test wells. Using a diluting loop or tulip, the meniscus of the test serum is touched (the loop should have been flamed, quenched in distilled water and blotted on absorbent paper). The loop is rotated in the well and then taken out and 0.025 ml of solution is delivered to the next well of the tray similarly. In this way serial dilutions of antiserum is done till the end of the plate. This is done in two sets. To one set of wells, antigen conjugated 1% SRBC suspension is added, one drop per well, while to the other set of wells, (as control), one drop per well of 1% SRBC suspension is added. The tray is covered with a sealer strip and the contents of the wells are mixed by shaking. Incubation is done at room temperature for 1 hour. Apart from various agglutination reactions set up as explained above, one drop of coated or uncoated cells are placed in separate wells to test for spontaneous agglutination.

Positive agglutination is seen when the cells form a continuous carpet on the base of the cup. If no agglutination has occurred, the cells fall as a tight button to the bottom of the V-shaped cup. If agglutination occurs with control cells, the antiserum must be absorbed to remove heterophile antibodies. If the coated or control cells agglutinate spontaneously, 1% normal serum is added to the buffer used.

Apart from haemagglutination tests, other particles such as bentonite or latex may be used as antigen carriers for agglutination tests. They have the advantage of not being antigenic. Latex beads provide a convenient carrier for antigens in agglutination tests. Sophisticated equipment has-been developed (PACIA, Technicon) which quantitates the degree of agglutination of the latex beads very accurately. This apparatus has led to a whole new field of immunoassay known as "Particle Counting Immunoassay" which has a sensitivity approaching that of radioimmunoassay. Coating of latex beads with antigens is not difficult, especially when using IgG or albumin as antigens, as these bind readily to polystyrene latex but many proteins, including other immunoglobulin isotypes bind less well. However, substitution of DNA groups into these proteins enables them to bind firmly to the latex. The best technique however is the coupling of latex particles covalently with carbodiamide. Coupling of antigens to latex is done by incubation of the latex with antigen at room temperature for 30 minutes, in glycine-saline. The latex is then washed by centrifugation and its remaining protein-binding sites are blocked using a solution of an irrelevant protein. The coated latex beads are stable at 4°C. In the latex agglutination test, doubling dilutions of the antiserum are prepared. To 25 ml of coated latex on a glass slide, 25 µl of the antiserum (of a particular dilution) is added and the side is rocked gently for 2 minutes. Agglutination can be read visually, illuminating the slide from the side, against a dark background.

Diagonostic Tests

Several commercial agglutination tests are available, and a few of them have been described below:

1. *WIDAL TEST*[44]

Widal Tube Test and Rapid Slide Test are used for detecting specific antibodies in serum of patients suffering from enteric fever. *In vitro* detection and quantitative estimation of specific antibodies can be done by this test. Stained *Salmonella* antigens (toxin) are inactivated, dyed, and mixed with suspensions of internationally recognised strains of bacteria. Antibodies produced in response to exposure to *Salmonella* will agglutinate bacterial suspension which carry homologous antigens (toxin). Readymade kits are available for the purpose and provide the following reagents:

- (i) Stained *Salmonella* antigen (toxin) *S. typhin* ''H''
- (ii) Stained *Salmonella* antigen (toxin) *S. typhi* ''O''
- (iii) Stained *Salmonella* antigen (toxin) *S. paratyphi* A (H)
- (iv) Stained *Salmonella* antigen (toxin) *S. paratyphi* B (H)
- (v) Positive control serum.

The kits also provide reusable glass slides and disposable mixing sticks.

Fresh and clear serum should preferably be used in the test and should not be heated or inactivated. three main tests are used generally, which are the rapid slide test, quantitative slide test, and quantitative tube test.

A. Rapid Slide Test

The procedure for this test is as follows:

1. Place one drop of undiluted test serum in each of the first four circles. (1 to 4) and one drop of positive control serum in each of the last two circles (5 and 6).
2. Place one drop of antigen (toxin) O, H, A(H), and B(H) in circles 1, 2, 3 and 4 respectively and O antigen in circle 5 and any one of the 'H' antigens [H, A(H) or B(H)] in circle 6.
3. Mix the contents of each circle with separate applicator stick and spread fluid over the entire area of the circle.
4. Rotate the glass slide for one minute and observe for agglutination.

If agglutination is visible within one munute, proceed for the quantitative slide test or tube test for quantitative estimation of the titre of the specific antibody.

B. Quantitative Slide Test

The rapid slide test is only a qualitative test and determines whether or not there is antibody to *Salmonella* antigens in the serum of the patient and types the antibody. Quantitation of antibody titre is done by the Quantitative slide test as follows:

1. Place 0.005 ml, 0.01 ml, 0.02 ml, 0.04 ml and 0.08 ml of undiluted serum in the 1, 2, 3, 4 and 5 circles respectively on the slide.
2. Add one drop of appropriate antigen (toxin) suspension which showed agglutination in rapid slide test to each of the above circles.
3. Mix the contents of each circle with separate applicator stick.
4. Rotate the slide for one minute and observe for agglutination.
5. The titre of antibody is the highest dilution of serum upto which there is clear agglutination.
6. Repeat the above steps with all the antigens (toxins) which showed agglutination in rapid slide test.

C. Quantitative Tube Test

This is another quantitative test for estimation of the titre of specific antibody as indicated by the rapid slide test. The procedure is as follows:

1. Take a set of 8 clean dry 10 x 75 mm test tubes for each serum sample under test.

2. Dilute each serum sample and set up the test as follows:

Tube	1	2	3	4	5	6	7	8
Dilution	1 : 20	1 : 40	1 : 80	1 : 160	1 : 320	1 : 640	1 : 1280	Saline control
N-saline (ml)	1.9	1.0	1.0	1.0	1.0	1.0	1.0	1.0
Serum (ml)	+ 0.1 un-diluted	+ 0.1 from tube 1	+ 0.1 from tube 2	+ 0.1 from tube 3	+ 0.1 from tube 4	+ 0.1 from tube 5	+ 0.1 from tube 6 (Discard 1 ml)	+ –
Antigen (drop)	1	1	1	1	1	1	1	1

3. Mix well and incubate at 37°C for 16 to 20 hours and examine for agglutination.

4. Repeat the above for all the antigens (toxins) that agglutinated in the rapid slide test.

5. The antibody titre is the highest dilution of serum showing clear agglutination.

The final dilutions of 0.08 ml, 0.04 ml, 0.02 ml. 0.01 ml and 0.005 ml in the quantitative slide test correlate approximately to the tube test dilutions 1 : 20, 1 : 40, 1 : 80, 1 : 160, 1 : 320 respectively.

Agglutinin titre greater than 1 : 80 is considered significant and usually suggestive of infection. The rising titre is considered as a definite evidence of infection.

2. *SYPHILIS RPR QUICK TEST*[45, 46, 47, 48, 49, 50, 51, 52]

The RPR (Rapid Plasma Reagin) test is a macroscopic non-treponomal flocculation test that is used to detect and qualify reagin antibody. When these antibodies are detected, it is a presumptive diagnosis for syphilis. The antigen (toxin) used in the kit is a modification of VDRL antigen (toxin) which contains microparticulate charcoal to enhance the visual difference between a positive and negative result. Rapid Plasma Reagin (RPR) test is a standard test for syphilis in which the cardiolipin-lecithin - cholesterol antigen is modified by the incorporation of choline chloride;

this modification permits serum to be tested without prior heating, unlike the VDRL Test. The kit provides the following reagents:

(i) RPR antigen (toxin) suspension, which is cardiolipin suspension-containing microparticulate charcoal

(ii) RPR control serum (positive)

(iii) RPR control serum (negative)

It also provides disposable test cards with 10 cells, 16 μl dispensing dropper needle for RPR antigen.

Plasma, heated or unheated serum may be used. The samples should be nonhaemolysed and free from contamination. Fresh serum samples may be stored at 2-8°C for 5 days or at –20°C for 4 weeks.

As in the Widal test, the test may be performed either qualitatively or quantitatively. In the qualitative test, after mixing a drop of the serum sample with a drop of the RPR antigen suspension the slide is placed on a mechanical VDRL Rotator operating at 100 rpm for 8 minutes and agglutination observed by reading the results macroscopically in direct light. Positive and negative control tests are run in parallel for comparison. A reactive result is indicated by large aggregates in the centre of the periphery of the test circle. A negative (non-reactive) result will display a smooth, even appearance with no aggregates visible while a weak reactive result is indicated by the presence of small aggregates around the edge of the circle.

Reactive Weak reactive Non reactive

In the quantitative test, the samples are diluted in physiological saline (0.9% NaCl) to give the dilutions 1 : 2, 1 : 4, 1 : 8, 1 : 16 and proceed as in the qualitative test. The last dilution step that contains macroscopic aggregates indicates the titre of the sample. If a 1 : 32 dilution gives a positive result the dilution series should extended to 1 : 64, 1 : 128, 1 : 256 and 1 : 512.

The RPR Quick test may give a false positive result which can be caused by diseases such as leprosy, lupus erythematosus, malaria, vaccinia infections and viral pneumonia. The final diagnosis should be based on a correlation of test results with other clinical findings.

3. *RF-AGGLUTINATING QUICK TEST*[53, 54, 55, 56, 57]

Patients with rheumatoid arthritis often develop antibodies to 'self IgG'. This antiglobulin antibody, known as Rheumatoid factor (RF) is readily detected by its ability to agglutinate latex particles coated with IgG.

The chemical significance of RF-determination consists in differentiating between rheumatoid arthritis, in which the rheumatoid factor has been demonstrated in the serum of approximately 80% of the cases examined and rheumatic fever in which the rheumatoid factor is almost always absent. The RF test is more frequently positive in active processes of greater duration than in diseases which are less active or are still in early stages. It is occasionally found in serum of patients with polyarteritis nodosa, systemic lupus erythematosus, hepatitis and certain other diseases. As in other agglutination tests, the test can be performed in a qualitative mode or as a semi-quantitative slide or tube test. The agglutinating reagent shows a sensitivity of IU/ml (slide test) of the International Reference preparation of rheumatoid arthritis sera. To detect the titre, serial dilutions of test serum from 1 : 20 to 1 : 10240 are made with 0.9% N-saline. The level of RF in the serum sample can be determined as follows:

RF in IU/ml = sensitivity in IU/ml (slide/Tube test) x titre (Slide/Tube Test).

4. *MINUTEX TEST FOR FIBRIN D-DIMER*[58, 59, 60]

Fibrin D-dimer is the characteristic end product formed by plasmin degradation of fibrin. The normal level of D-dimer in plasma is < 200 ng/ml. Elevated levels of D-dimer (300 to > 10,000 ng/ml) are indicative of on-going fibrinolysis and are found in various pathological conditons including deep venous thrombosis (DVT), pulmonary, embolism (PE), disseminated intra vascular coagulation (DIC), and pregnancy complications. The minutex D-dimer test is a latex agglutination assay for rapid detection of elevated levels of fibrin D-dimer. It uses the monoclonal anti-D dimer antibody MA-8D3 coated on latex particles and the result can be read after 180-200 seconds of mixing the serum sample with the latex suspension.

5. *BETA VISIPREG TEST FOR PREGNANCY DIAGNOSIS*

Latex particles may also be used in inhibition assays and a commonly used pregnancy test is based on this principle. Human chorionic gonadotropin (HCG) is present in the urine of pregnant women after conception and can be detected by its capacity to inhibit the agglutination of HCG-coated latex by an anti-HCG serum.

Beta Visipreg offers a high sensitivity (0.5 IU/ml), facilitating early pregnancy diagnosis. It uses a blend of β-HCG specific monoclonal antisera which offers 100% specificity and a very high degree of accuracy. It is a simple test requiring only two minutes and since it is an *in vitro* economic and reliable test, it is preferred for pregnancy diagnosis.

Complement Fixation Test (CFT)[61, 62, 63]

This is a sensitive reaction used in serology for the detection of either antigen or antibody and is used for the diagnosis of many bacterial and viral diseases, including syphilis. As dealt with in Chapter 1, antibodies can 'fix' or 'activate' complement when they combine with their homologous antigens. IgG (subclasses 1, 2 and 3) and IgM can bring about complement fixation *via* the classical pathway; IgG 4 and IgA (subclasses 1 and 2) can activate the alternate pathway.

A complement fixation test (CFT) is a term that refers to any range of sensitive, *in vitro* tests in which complement fixation is used to indicate the presence or absence of or quantity of a specific antigen or a complement-fixing antibody. The principle of CFT is that when antigen and antibody are allowed to interact in the presence of complement, the amount of complement used up (which in turn is a measure of the amount of antigen-antibody reaction) is estimated indirectly by the addition of a haemolytic system; the proportion of erythrocytes lysed indicates the amount of free (unfixed) complement remaining.

The test involves two stages: stage I is the binding or fixation of complement if certain antigen-antibody reactions occur, and stage II is the detection of residual unbound complement, if any, by its haemolytic action on the sensitized erythrocytes subsequently added. In the first stage, either the antigen or the antibody must be supplied as a reagent with the other of the pair as the test unknown. Fresh guinea pig serum is normally used as the complement source. Sheep erythrocytes which are coated with their corresponding antibody (amboceptor or haemolysin) are used in the second stage. Such erythrocytes form what is known as haemolytic system. It is a suspension of 'sensitized' erythrocytes, i.e. erythrocytes which have been exposed to a subagglutinating dose of antibodies homologous to their surface antigens. The system exhibits no haemolysis in the absence of complement, and most complete haemolysis (90% of the erythrocytes are lysed) when exposed to the maximum concentration of complement, used in the CFT. In CFT, a lack of haemolysis denotes that the complement is bound and this result is termed a positive CFT and is indicative of presence of the antigen or antibody being tested for.

In practice, serial dilutions of the inactivated serum (serum which has been heated to 56°C for 30 minutes to inactivate the complement it contains) are prepared, and known, fixed quantities of complement and antigen are added (if antobody has to be detected and quantified in the serum) to each dilution. The quantity of complement added must be sufficient to cause lysis of at least 90% of the erythrocytes in the haemolytic system when no complement-fixation has occurred. Incubation is carried out for 18 hours at 4°C (since complement components may be heat liable). A fixed volume of the haemolytic system is then added to each dilution and the whole is incubated for 15-30 minutes at 37°C. The test dilutions are then left at 4°C to permit non-lysed erythrocytes to settle. In any given dilution, the extent of haemolysis will be inversely proportional to the extent of complement fixation in that dilution. If maximum haemolysis is observed in the lowest serum dilution (i.e. maximum serum concentration) the test is negative — i.e. the serum contains no detectable antibody. In a positive test, no haemolysis occurs in (at least) the lowest serum dilution. In all CFTs, controls are essential, e.g. to check that the test system is functioning correctly and to preclude false - positive interpretations based on the effects of anticomplementary substances, which are any substances (eg. zymosan), or any system (e.g. pre-formed antigen-antibody complexes) other than the specific combination of antigen with antibody which brings about complement fixation.

Microcomplement fixation is a method for comparing the amino acid sequence in a given reference protein (e.g. an enzyme) with that in the same protein derived from a different strain or organism; the method is commonly used in microbial taxonomy. Essentially, the protein under test is examined for its ability to fix complement when mixed with an antiserum raised against the purified reference protein. The quantitative difference in the complement-fixing ability of the test and reference proteins is taken as a measure of the difference between the amino acid sequences of the two proteins. The results are usually given in Imd units; 5 Imd, units represent 1% difference between the amino acid sequences.

Complement fixation can be demonstrated in immunodiffusion tests. An antigen is diffused toward its antibodies in a double diffusion plate through agar containing sheep erythrocytes and complement. Later the gel is flooded with haemolysin, and erythrocytes throughout the gel lyse except in the zone where antigens and antibodies have combined and fixed complement[64].

Complement Fixation tests find applications both in treponomal and non-treponomal tests for syphilis.[65] Treponomal tests are those in which treponomal antigens or intact cells are used to detect specific anti-treponomal antibodies which occur in the serum of patients suffering from syphilis. Tests in which a non-treponomal antigen is used to detect or quantify non-treponomal antibodies (reagins) which occur in the serum of syphilis patients are called standard tests for syphilis (STS) or non-treponomal tests. The RPR agglutination test, the VDRL agglutination test, etc., for example, are standard tests for syphilis. A standard test for syphilis known as the Wassermann Reaction (WR) is carried out as a CFT in which the antibody (reagin, 'Wassermann antibody') undergoes a complement - fixing reaction with the cardiolipin-lecithin-cholesterol antigen. Treponomal tests are more specific than the standard tests for syphilis and are generally used as confirmatory tests (e.g., when the results of STS are equivocal). However as they remain positive for long periods after effective therapy, treponomal tests are less suitable than STA for assessing the progress of therapy. Treponomal tests include the FTA-ABS Test, TPHA Test, Reiter protein complement fixation test (RPCFT) and the TPI test. The FTA-ABS Test involves fluoro immunoassay principles and will be dealt with later. The TPHA test *Treponema pallidum* haemagglutination test involves passive haemagglutination in which an absorbed serum is tested for its ability to agglutinate tanned erythrocytes sensitised with antigens from the Nichol's strain of *Treponema pallidum.* The TPI test (*Treponema pallidum* immobilization test) is a treponomal test in which antibodies specific to *T. pallidum* and some other treponemal species are detected. Essentially, the test depends upon the immobilisation of living (motile) cells of *T. pallidum* in the presence of specific antibodies and complement. To the serum under test is added a standardised suspension of *T. pallidum* and a volume of fresh guinea-pig serum as a source of complement.The whole is incubated anaerobically for 18 hours at 37°C and examined microscopically. In a positive reaction, a specified proportion of treponemes is immobilised. In Reiter protein CFT, the Reiter antigen associated syphilis (the reference strain of *Treponema phagedensis,* contains significant amounts of antigen(s) common to a number of species of *Treponema,* including *T. pallidum* and these form the Reiter group Treponomal antigens) can be diagnosed by a CFT.

CHAPTER 4

Isotopic and Non-isotopic Labelled Immunoassays

Introduction

Immunoassays can be broadly classified as labelled and nonlabelled. The methods we have discussed so far do not involve any labelling of the reactants and are based on a direct consequence of the antigen-antibody reaction which is visualised. This method of detecting the antigen or antibody does not work when the concentration of the reactants(s) is below that which is required to allow the visualisation of the antigen antibody complex as a precipitate or allow agglutination, etc. Also, the specificity of non labelled immunoassays is much less than of labelled assays. These observations led to the development of the more versatile, specific and sensitive techniques of labelled immunoassays in which the final read-out of the reaction involves a reagent conjugated with an appropriate label. A wide variety of 'labels' or 'tags' are available based on which labelled immunoassays are classified as: Radioimmunoassay, Enzyme-linked immunosorbent assay, Fluoroimmunoassay, Luminiscent immunoassay etc. Each method has its own limitations, and advantages and the choice of the method depends on the reactants, the sensitivity and specificity required, health hazard considerations and economic factor.

Radioimmunoassay (RIA)

Radioimmunoassay is a highly specialised technique with an enormous diversity of applications. It is rapid, sensitive and precise. In RIA, the concentration of an unknown antigen is determined by measuring its ability to compete with a fixed amount of radiolabelled antigen for a limiting quantity of antibody. The antibody may either be immobilised on an insoluble support or in solution. Similarly, RIA for antibody can also be performed. Depending on whether the test is done using an immobilised reactant or in solution, RIA can be classified as:

1. Soluble Phase RIA, and
2. Solid Phase RIA.

Soluble phase RIA can again be done for either antibody or antigen. If a reasonable excess of labelled antigen is added to an antiserum most of the antibodies of moderate affinity will be complexed and precipitation of the complex and measurement of the label will give an estimate of the antigen-binding capacity of the serum. This is schematically shown below:

*Ag +	Ab ⟶	*Ag +	*AgAb ⟶	*AgAb
Excess Labelled Antigen	Antiserum	Free antigen	Soluble Complexed antigen	Precipitated Complex

↑ 50% Ammonium sulphate
OR
2% Polyethyleneglycol
OR
Anti-Immunoglobulin
OR
Staphylococcal Protein A.

By using antibodies to different immunoglobulin classes, it is possible to determine the distribution of antibody activity among the classes. For example, addition of a radioactive antigen to human serum followed by a precipitating rabbit anti-human IgA would indicate how much antigen had been bound to the serum IgA. Soluble phase immunoassay for antigen is also known as the classical RIA for antigen. The principle of the assay is that the binding of radioactively lebelled antigen to a limited fixed amount of antibody can be partially inhibited by the addition of unlabelled antigen and the extent of this inhibition can be used as measure of the unlabelled material added, this is illustrated in Fig. 4.1. Methods vary in the means used to separate free antigen from that bound to antibody. These may be

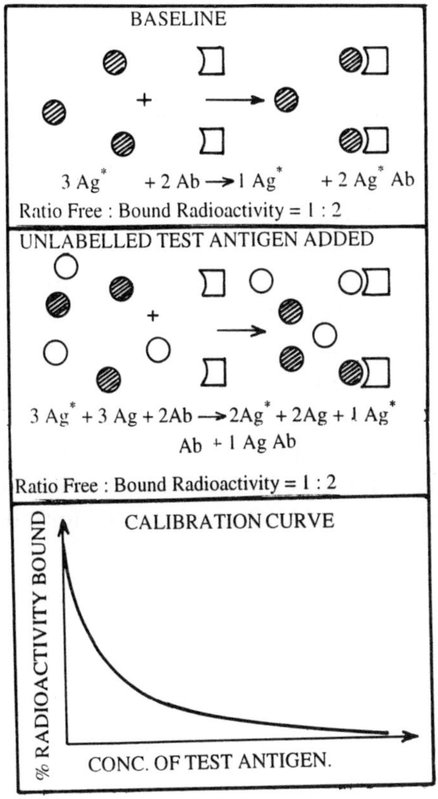

Fig. 4.1. Principle of RIA. Baseline indicates a control reaction without added test antigen. O = unlabelled antigen, ● = labelled antigen, ⊐ = antibody

precipitation with 50% ammonium sulphate, 2% polyethylene glycol (PEG), anti-immunoglobulin or *Staphylococal Protein* A. If the antigen can be labelled to a high specific activity, very low concentrations down to 10^{-12} g/ml level can be detected by RIA. This allows the quantitation of most of the protein hormones by this technique. Other applications include the assay of carcinoembryonic antigen, *hepatitis* B (Australia) antigen, and smaller molecules such as steroids, prostaglandins and morphine-related drugs. However, these assays have the disadvantage that they cannot distinguish active protein molecules from biologically inactive fragments which still retain antigenic determinants. This should be considered when using the RIA for quantitation of protein hormones.

In RIA, standard curves are constructed by varying the amount of unlabelled antigen (x) added to fixed amounts of labelled antigen and diluted antiserum. The measured response (y) is the ratio of labelled antigen bound in the presence of unlabelled antigens. Straight plots are obtained if logit Y or log $(y/(1-\alpha y))$ is plotted against log x, that is, logit Y = a + b log x, where b is the slope and a is the Y intercept. (Fig. 4.2). If plotting the varying concentrations of an unknown yields a curve parallel to that obtained in plotting varying concentrations of standards, it is likely that unknown reacts with the same antibody as the standard. Therefore, the unknown probably contains the same antigen. Relative concentration of antigen can then be estimated by comparing its effect on Y with that of standard. An unknown antigen that reacts with the same antibody, but at different affinity, will not yield a parallel curve. In other words, cross-reactive antigens (similar but not identical epitopes are present) will not yield a parallel curve. Lack of parallelism, therefore, is suggestive of measurement of an antigen other than the specific antigen under investigation. Figure 4.2 shows the straight line plot obtained in RIA upon plotting logit of the ratio of labelled luteinising hormone-release hormone (LHRH) bound in presence to that bound in absence of unlabelled LHRH against the log of concentration of a labelled LHRH. When an extract of arcuate nucleus (ARC) is substituted for unlabelled LHRH in

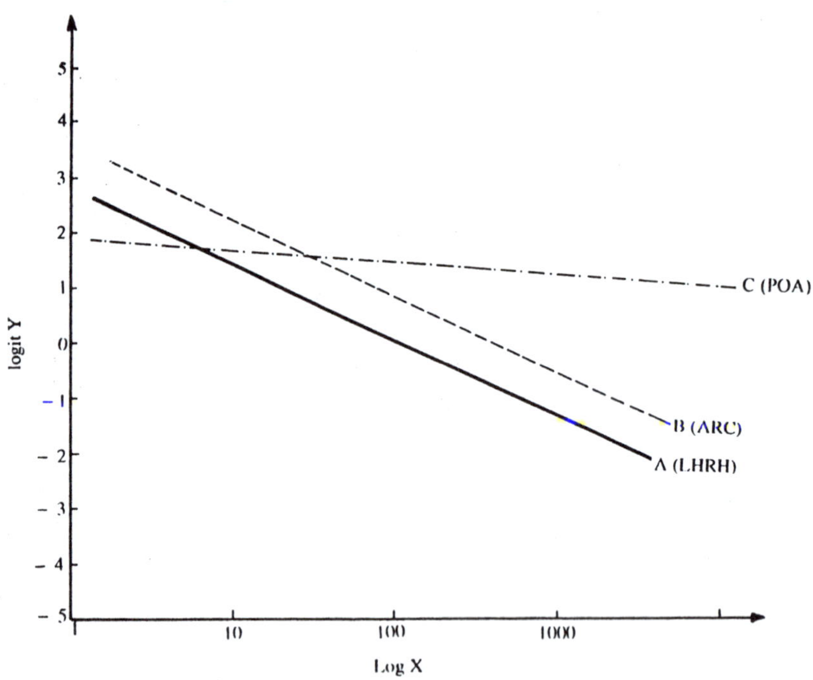

Fig. 4.2. RIA standard curves obtained by plotting logit Y vs log X.
A. Standard curve for LHRH
B. Curve with an unknown sample from arcuate nucleus (ARC)
C. Curve with an unknown sample from preoptic area (POA)
Source: Immunocytochemistry, Ludwig. A. Sternberger; John Wiley & Sons, 1986

the same experiment, a straight line with similar slope is obtained, suggesting the presence in the extract of a substance similar or identical in structure to LHRH. However, when a hyphothalmic extract from the preoptic area (POA) is used instead of unlabelled LHRH, a nonparallel line is obtained indicating that the principle in the preoptic area is crossreactive with, but not identical to LHRH.[66]

Solid Phase RIA differs the soluble phase immunoassay in that one of the reactants is immobilised on a suitable solid matrix. The antibody content of a serum can be assessed by the ability to bind the antigen which has been insolubilised by physical adsorption to a plastic tube or micro-agglutination tray with multiple wells. The bound antibody is then measured by the addition of a radiolabelled second antibody, which is an antibody raised to the first antibody in another species. This binds to the complex formed in the first step and after washing the excess antibody that is unbound, measurement of radioactivity in the tube or well will give an indication of the concentration of the antibody. Solid phase RIA is used in the diagnosis of *Lupus Erythemtasus* in which autoantibodies are produced. DNA autoantibodies are quantitated by using microwells coated with DNA and performing the test using the patient's serum and ^{125}I labelled purified rabbit anti-human IgG. After adding the second antibody, excess unbound reagent in the tube can be rinsed out and the radio-activity in the tube will be a measure of the autoantibody content of the patient's serum[67]. Another application of solid phase RIA is the radio allergosorbent test (RAST) for IgE antibodies in allergic patents. The antigen (the allergen, say pollen extract) is covalently coupled to an immunoabsorbent which is a paper disc in this case. This is then treated with the patient's serum followed by the addition of labelled anti-IgE. Solid Phase RIA for antigen is known as the Immunoradiometric assay for antigen. Here, an antibody to the antigen to be measured is coated on the surface of a solid support such as plastic and to this the sample is added. Then excess of radio labelled antibody which recognises an epitope on the antigen different from that recognised by the coated antibody is added. Thus, in this assay the labelled reagent is added in excess. Bound antigen is thus measured. By using labelled antibodies which recognised different parts on the antigen, the specificity of this technique can be increased. An example is in the radio immunosorbent test (RIST) for IgE which uses Rabbit-anti-IgE coupled to microcrystalline cellulose by cyanogen bromide and labelled anti-IgE in excess.

Manufacture and Use of a RIA Kit

The manufacture of a RIA kit involves the following:

1. Antiserum production

2. Label production
3. Phase separation
4. Immunoadsorption (for solid phase RIA)

1. Antiserum Production

Depending on the situation, either polyclonal or monoclonal antibodies may be used. Polyclonal antibodies are produced by immunization. The antigen is made sufficiently immunogenic by covalent coupling of the hapten to the carrier. Monoclonal antibodies may be produced by the hybridoma technique.

2. Label Production

Antigen or antibody depending on the nature of the assay has to be radiolabelled. A high-specific activity label is a must for the development of a sensitive RIA. In practice, either a multiple tritium substituted molecular species of the substance to be measured or an analogue of the substance to which a ^{125}I or ^{131}I tag has been attached, is used. Radiolabelling by iodination seldom presents difficulties in RIA of protein and large peptide molecules. However, problems do arise with low molecular weight compounds due to differences in the immunoreactivity of the labelled product and the original hapten.[68] Both ^{3}H and ^{14}C-labelled materials which have half-lives of several years, have similar, though not always identical immunoreactivities to the native hapten, and do not suffer radiation damage unless they are of extremely high specific activity. Their main disadvantages, especially those based on ^{14}C are that they have low specific activity, thereby reducing the sensitivity of the RIA to unacceptably low levels. In order to overcome these difficulties, labels incorporating shorter half-life γ-emitting radio nuclides, such as ^{125}I and ^{75}Se, have been frequently used in development of a RIA. Several techniques are available that allow the introduction of a label without a significant alteration in the immunological properties of the labelled molecues. These include:

1. Chloramine T method for iodination of proteins
2. Iodogen method for iodination of proteins
3. Biosynthetic labelling of hybridoma derived antibody
4. Cell surface iodination by lactoperoxidase technique.

3. Phase Separation Techniques

A RIA kit should also provide a method of separating the free antigen from the antibody-bound fraction of the antigen without disturbing the antigen-antibody equilibrium. Contamination of one fraction with label from the other results in erroneous values for the amount of antibody bound and hence, inaccuracies in assay.

There is no single method of phase separation that is suitable for all assays and several different types of separation procedures have been developed, mainly on physico-chemical and immunological concepts. Chromatographic techniques are generally too cumbersome for routine use involving simultaneous analysis of many samples, but may be extremely useful in the initial stages of new assay development. Adsorption methods are usually favoured since they are simple, rapid and cheap. Usually the free antigen fraction is adsorbed on to the material selected, leaving the bound fraction in solution. Particulate charcoal[69] is the most widely used adsorbent, especially for the assay of small molecules, but is rarely used in untreated form, since it is capable of binding not only to free antigen but also to a lesser extent to antibody-bound antigen. In order to try and alleviate this difficulty, particulate charcoal is generally coated with dextrin before being used in RIA, the molecular weight of the dextrin employed being dependent on the molecular weight of the antigen involved. Dextrin possibly acts as reverse molecular sieve, permitting the small 'free' antigen molecules to penetrate into, and hence be adsorbed by the inner core of the charcoal, whilst excluding the larger antigen-antibody complexes. This theory of action has, however, been disputed.[69] As a safeguard against the 'stripping' effect of charcoal on the antigen-antibody complex, it is advisable to perform separations at 4°C and for as short a time as practicable.[70] An alternative to the adsorption of free antigen is the precipitation of the antibody-bound fraction. This can be achieved by physicochemical (salt precipitation) or immunological(double antibody) means. The latter, though time-consuming and expensive, is generally advantageous when only low avidity antiserum are available. In other methods, insoluble supports to which either the first or the second antibodies are covalently attached, are used. The solid phase used in such systems can be almost any insoluble material capable of making suitable links and include cellulose, high molecular weight dextrins and even the plastic tubes in which the assays are preformed. The use of such solid phase antibodies can greatly simplify assay procedures provided they retain their avidity and do not leach off the matrix.

4. Immunoadsorption

In the manufacture of a solid phase RIA kit, the production of the immunoadsorbent (insolubilised form of antigen) is important. Antigen is usually bound covalently on a matrix to produce an immunoadsorbent.

Cyanogen bromide-activated sepharose reacts with amino groups of antigens. some haptens, such as luteinizing hormone-releasing hormone (LHRH), do not contain amino groups. In such case, the hapten can be immobilised by use of a spacer, such as RNAse or any other protein. LHRH is first coupled with RNAse via its tyrosine groups by bis-diazotised benzidine, and product is then reacted with activated sepharose via the amino groups of RNAse.[71]

The use of a RIA kit involves the following:

1. Obtaining a standard curve
2. Blank and control parameters
3. Accuracy and precision
4. Calculation of results.

1. Obtaining a Standard Curve

The range of standards included in the standard curve is determined by the concentration of the antigen expected in the samples undergoing analysis. Preferably the range should be such that the samples do not need to be diluted by complicated multiple dilutions so that diluting errors are minimised. When sample concentrations are likely to lie within a wide range, several dilutions of the sample may be necessary. In such cases, it may be advantageous to construct two standard curves with different sensitivities. Doubling dilutions of the standard should be avoided since any error in pipetting then becomes cumulative.

2. Blank and Control Parameters

In order to monitor the performance of a RIA over a prolonged period, certain 'Blank' and 'Control' tubes must be included in each assay. Any change in the antibody binding reactions can be detected by incubating labelled antigen with an excess of antibody. Reduction in the 'maximum bound' value can be due to various factors, including the loss of immunoreactivity of the labelled antigen, interference with its binding to the antibody by some non-specific factors such as pH, ionic strength, etc. in the incubation mixture. Nonspecific binding of the label, either to the walls of the assay tube or to buffer constituents should be determined by incubating the label alone in assay buffer. An increase in nonspecific binding may indicate deterioration of the labelled antigen and indicate the need for its replacement. Control samples of known compositions should always be included in each assay. In order to minimise random errors in RIA, it is common practice to set up standards and samples in duplicate.

3. Accuracy and Precision

Although the actual procedure for setting up a RIA is simple, once optimum conditions have been established, great care must be taken to minimise errors in pipetting and dilution. Automatic pipetters and dispensers are preferable, especially when a large number of samples have to be handled. Radioactive counting errors can be minimised by adding at least 10,000 counts to the assay tubes. If counting time and space permits, errors may be further reduced by counting both the bound and free fractions.

4. Calculation of Results

There are several alternative methods of constructing the dose-response (standard) curve from which the amount of antigen in the unknown samples can be computed. The amount of radioactivity in the antibody-bound fraction (B) is often corrected by substracting the apparent nonspecific binding (BSB) from the observed figure before constructing the curve. Statistical analyses may be simplified by the use of logit-log plot in which logit (B/B_o) is plotted against log (Ag), where B is the amount of radioactivity bound in the presence of a certain antigen concentration (Ag) and B_o is the amount bound in the absence of the standard. The advantage of this type of plot is that it often yields a linear relationship (Fig. 4.2) permitting its use in automatic data processing systems. Logit-log linearity requires the dose response curve to conform to sigmoidal shape. This is not always the case, because some logit-log plots which are linear over much of the antigen concentration range, may deviate from linearity at the extremities and errors occur if sample antigen values are calculated from points beyond these limits.

The sensitivity of RIA is determined by the avidity of the antiserum and can be calculated in many ways. A 10% fall in binding from that in the absence of added antigen is often taken as the limit of detection. Antibodies with avidity constants of 10^8-10^{10} are sufficiently avid for use in RIAs. Also if a RIA is intended for determination of serum-drug levels (therapeutic drug monitoring), it is preferred since at this avidity the antibodies have the capacity to overcome the plasma-protein binding effects exerted by many drugs.

A Soluble phase RIA kit has the following components:

1. Antigen standard
2. Antiserum standard
3. Radiolabelled antigen

4. Antigen free human serum
5. Control sera: positive and negative
6. Assay buffer
7. An agent for separating free and bound antigen [generally polyethylene glycol (PEG)]

A typical example of such a kit is the one used for assaying the levels of thyroxine in human serum. Its main effect is to increase metabolic activity and decreased thyroxine levels have been seen in hypothyroidism, chronic thryroiditis and cretinism. Elevated T4 levels have been associated with hyperthyroidism, hepatitis, pregnancy and exogenous administration of certain drugs like salicylates, barbiturates, dilantin, etc., elevate or depress thyroxine levels. 99.5% of the thyroxine circulating in the blood is bound to specific proteins collectively known as thyroxine binding globulins (TBG). This protein binding protects the thyroid hormones from metabolism and excretion resulting in the long half life (6-7 days) of thyroxine in circulation. Thyroid disorders are among the most common endocrine disorders encountered in clinical practice. In the test, unlabelled endogeneous thyroxine competes with raidio-labelled hormone for the limited binding sites on a monoclonal antibody against thyroxine. At the end of incubation, the thyroxine bound to antibody and free thyroxine are separated by the addition of 22% PEG. The amount bound to the antibody in tube is compared with the values of known thyroxine from which standard thyroxine concentration in the patient's sample can be calculated. 8-aniline 1-napthalene sulphonic acid (ANS) is used in this kit for displacing thyroxine bound to TBG. The test requires 10μl of serum volume and covers a sample range of 0-200 ng.ml. Assay buffer used is 0.1% gelatin in 0.14M THAM (Trishydroxymethyl aminomethane), pH 8.6 ^{125}I labelled thyroxine is used as labelled antigen and the radioactivity is measured using well type gamma scintillation counter. RIA finds numerous applications in the *in vitro* diagnosis of many endocrine abnormalities. A number of protein hormones can be quantitated by RIA. The applications of RIA are listed in Table 4.1. However, the technique has some disadvantages as well. Because of health hazards and the deterioration of reagents through radiation damage, other types of labelled immunoassays have gained popularity.

Non-Isotopic Imunoassays

Although RIA permits quantification of a great variety of biologically active compounds, it does have certain drawbacks. Beta-emitting isotopes, for example, necessitate the use of expensive scintillants and liquid scintillation counters, and often need long count time making the assays slow and tedious. The radioactive and organic wastes must be disposed off safely. Compounds labelled with gamma-emitting isotopes, e.g., ^{125}I, ^{131}I and ^{75}Se, have relatively short shelf-life and create slight health hazards during their preparation. This has not prevented the fear of such dangers being exaggerated beyond all reasoning and the virtual banning of RIA in some parts of the world and serious restriction being put on its use in most others. For all these reasons, alternative labels have been investigated in an attempt to retain the advantages of specificity and sensitivity of immunoassay without using radiolabels. Labels that have been investigated include enzymes, fluorescent groups, stable free radicals, metals, erythrocytes and bacteriophages, all of which can, like radioisotopes, be attached either to the hapten or to the antibody. A potential advantage of many, alternative labels over radioisotopic labels, in immunoassay is their amenability to homogeneous assays, i.e., those where it is not necessary to separate antibody bound from free hapten. This opportunity does not exist for RIA since binding of antibody to an isotopically-labelled antigen does not affect the radioactivity of the isotope.

Table 4.1 Substances measured by Radioimmunoassay

Peptide hormones	Nonpeptidal hormones	Nonhormonal substances
Pituitary hormones	Thyroidal hormones	Drugs
Growth hormone	Triiodothyronine	Cardiac glycosides
Adrenocorticotropic hormone (ACTH)	Thyroxine	Morphine
Melanocyte-stimulating hormone (MSH)	Prostaglandins	LSD
α-MSH	Steroids	Barbiturates
β-MSH	Aldosterone	Chlorpromazine
Glycoproteins	Corticosteroids	Nicotine
Thyroid-stimulating hormone (TSH)	Estrogens	Penicillin and gentamicin
Follicle-stimulating hormone (FSH)	Androgens	D-tubocurarine
Luteinizing hormone (LH)	Progesterones	Cyclic nucleotides
Prolactin		Enzymes
Lipotropin (LPH)		C_2 esterase
Vasopressin		Fructose 1,6-diphosphatase

Peptide hormones	Nonpeptidal hormones	Nonhormonal substances
Oxytocin		Plasminogen and plasmin
Chorionic hormones		Chymotrypsin
Human chorionic gonadotropin (HCG)		Trypsin
Human chorionic somatomammotropin (HCS)		Carbonic andydrase isoenzymes
Pancreatic hormones		Virus
Insulin		Australia antigen (hepatitis B antigen)
Glucagon		Tumor antigens
Calcitropic hormones		Carcinoembryonic antigen
Parathyroid hormone (PTH)		α-Fetoprotein
Calcitonin (CT)		Serum proteins
Gastrointestinal hormones		Thyroxin binding globulin
Gastrin		IgG, IgE
Secretin		Properdin
Cholecystokinin-pancreozymin (CCK-PZ)		Fibrinogen
Vasoactive intestinal polypeptide (VIP)		Anti-Rh antibodies
Gastric inhibitory polypeptide (GIP)		Other
Vasoactive tissue hormones		Intrinsic factor
Angiotensins		Rheumatoid factor
Bradykinins		Hageman factor
		Folic acid
		Neurophysins
		Staphylococcal β-enterotoxin

Source: *Mc Graw-Hill Encyclopaedia of Science & Technology*, Vol. 15, 1987.

Enzyme Immunoassay (EIA)

As the name indicates, the label used in EIA is an enzyme. EIAs of several different types have been used for quantification of several clinically important substances like hormones, drugs, bacterial and viral surface antigens, antibodies, tumor markers, receptors, etc. EIAs can be broadly classified as substitutional assays and homogeneous EIAs.

As the name indicates, the substitutional assays involve the substitution of an enzyme for a radiolabel in an already well established RIA procedure. Homogeneous EIAs on the other hand are novel and utilize the enzyme label with special physical and chemical properties.

Two types of homogeneous EIAs will be described here. These are enzyme multiplied immunoassay technique (EMIT) and the Ames TDA. The principles of the EMIT technique developed by the Syva company depends upon the fact that binding of antibody to an appropriate enzyme labelled substance, prevents its enzymatic activity[72], either by sterically preventing the access of the substrate to the catalytic site or by inducing conformational changes in the enzyme which prevent catalysis.[73] Free antigen in the medium competes with enzyme-labelled antigen for antibody with the result that the enzymatic activity is uninhibited and proportional to the amount of free antigen originally present. Because each enzyme molecule is capable of catalyzing the conversion of an extremely large number of substrate molecules from an optically inert to an optically absorbent (or even fluorogenic) substance, the EMIT system is at least theoretically extremely sensitive. In reality, the sensitivity of the method is limited by the other constituents, notably the antibody-antigen binding characteristics and the presence of basic background noise generated by the reagents themselves. This technique is of great use in therapeutic drug monitoring for the measurement of plasma levels of various clinically important drugs such as digoxin, carbamazepine, antibiotics, antiepileptics. etc.

A second type of homogeneous EIA utilizing a totally different principle was described by Burd and coworkers and is commercially available as the Ames TDA system. In this type of assay, the antigen is covalently bound to the substrate of an enzyme in a way that renders it unavailable for hydrolysis when the antigen-substrate complex is bound to antibody but freely available when it is not. If in addition, the free substrate, and not its conjugate is fluorogenic, then the amount of fluorescence developed during the reaction will depend upon the quantity of substrate-antigen conjugate displaced by unlabelled antigen from the limited number of antibody binding sites available in the reaction mixture. In this type of assay, the number of fluorescent molecules available for measurement is equal to the number of molecules of antigen to be measured and imposes limitations upon even the theoretical sensitivity of the assay. Ames TDA is also used in therapeutic drug monitoring.

Enzyme-Linked Immunosorbent Assay (ELISA)

The substitutional EIAs are of much greater clinical significance than the homogeneous EIAs. Substitutional EIAs are otherwise referred to as Enzyme-Linked Immunosorbent Assay (ELISA) when they are performed using a solid phase technique. Competitive enzyme-immunoassays are analogous to classical RIAs in which an enzyme-labelled antigen competes with unlabelled antigen for binding to limited quantity of specific antibody sites. Free antigen is separated from antibody bound antigen and the enzyme activity in either fraction measured or related to the concentration of unlabelled antigen. In the immunoenzymometric assay, the antigen to be measured is permitted to react with an excess of labelled antibody. After incubation excess solid phase antigen is added to the mixture and binds any antigen-free labelled antibody present. Following separation of the solid phase enzyme activity is measured in either the solid phase or the supernatant. The amount of enzyme activity present in the supernatant is directly and stoichiometrically related to the amount of antigen originally present in the sample. This method is analogous to the immunoradiometric system for antigen assay. The sandwich method of ELISA is analogous to the two-site immunoradio-metric system and requires that antigen has at least two immunogenic binding sites. The solid phase is coated with the antibody. The sample is added and after incubation another antibody raised against another determinant of the antigen (but enzyme labelled) is added. The amount of enzyme activity associated with the solid phase is proportional to the amount of antigen present in the sample. This type of ELISA is frequently used in the quantitation of several cell surface antigens, antibodies, receptors, and tumor markers, for such solid-phase assays, polystyrene beads, red cells, agarose beads, etc. can be used as the support but it is more convenient to use a plastic test tube or microtitreplate as the carrier for the antigen or antibody.

The important step involved in making a ELISA kit is the conjugation of enzyme to the antibody. The enzyme used frequently is Horse Radish Peroxidase. The conjugation of Peroxidase and IgG can be done in many ways, which include:

1. Conjugation of peroxidase to immunoglobulin with glutaraldehyde
2. Conjugation of peroxidase to Fab hinge regions with maleimide
3. Periodate conjugation technique.

Conjugation with glutaraldehyde is a two-step technique. Glutaraldehyde is a dialdehyde. In the first stage, peroxidase is reacted with excess glutaraldehyde. About two amino groups in peroxidase react, each with one arm of glutaraldehyde. In the second stage, conjugated peroxidase still possessing one reactive aldehyde group on each of its two reaction sites is mixed with IgG. Ideally, monomeric products consisting of conjugates of one peroxidase and one IgG are formed. Polymeric by-products, however, interfere. If instead of peroxidase, other enzymes that usually possess more reactive amino groups are used for conjugation, a higher number of aldehyde groups are available for the second stage and polymer formation is increased. Also, if instead of glutaraldelyde, the aldehyde formed by oxidation of the carbohydrate moiety of peroxidase is used, the second step of the reaction will be more effective, and again the amount of interfering polymer will be increased. The amount and degree of polymer formed differs within the same procedure carried out at different times. In general, procedures that yield largely monomeric conjugates are low in yield and leave much interfering nonconjugated antibody in solution, while procedures that are high in yield result in excessive polymer formation. Some of the enzyme and some of the antibody reactivity may be strictly hindered in the interior of polymers. Polymers must be removed before a conjugate is to be used for the development of a good ELISA kit. This can be done by gel fillration using Ultragel AcA44. Unconjugated antibody competes with conjugate by blocking the epitopes in the sample. Furthermore, unconjugated antibody may have a higher affinity for antigen than the conjugated antibody. Hence, the unconjugated antibody has to be separated. Both immunoglobulin as well as peroxidase contain carbohydrate, but the proportion of carbohydrate to protein is six to nine times higher in peroxidase. The lectin concavalin A, binds these proteins, and therefore, concavalin A-sepharose affinity column removes peroxidase conjugated antibody as well as unconjugated antibody and peroxidase from solution. Arends[75] made use of a differential desorption of these proteins from the column by elution with 0.1M methyl-alpha-D-glucopyranoside which competes with the absorbed proteins for carbohydrate. Unconjugated immunoglobulin is released from the column first and good separation is obtained from the enzyme-conjugated immunoglobulin and the trailing unconjugated peroxidase.

Another conjugation technique utilises the carbohydrate moiety of peroxidase which is not needed for enzymatic activity. The procedure is done is two stages. In the first stage, the carbohydrate is oxidised to aldehydes using a reagent such as periodate. In the second stage, the aldehyde groups are allowed to react with immunoglobulin which involves Schiff base formation with amino groups in immunoglobulin. In order to avoid self coupling of peroxidase by reaction of its aldehyde group with its protein amine, the amino groups are blocked by fluorodinitrobenzene prior to the first stage. However, this step may be unnecessary if oxidation is carried out at a low pH. At the end of the second step, sodium borohydrate is used to stabilise the Schiff base. In the actual procedure, 0.2 ml of 0.1M sodium periodate are added under stirring for 20 minutes to 4 ml peroxidase dissolved in 1ml of distilled water. Following dialysis against 0.001M sodium acetate buffer, (pH 4.4) at 4°C, 20 ml of sodium carbonate buffer, (pH 9.5) are added and the pH is further raised to 9 to 9.5. The solution is added to 8 mg affinity purified antibody in 0.1M sodium carbonate buffer, pH 9.5. After stirring for 2 hours at room temperature, the reaction is stopped with 0.01 ml of freshly prepared sodium borohydrate (4 mg/ml) and the mixture is left at 4°C for 2 hours.[76]

Conjugation of peroxidase to Fab hinge regions with maleimide is a technique in which single peroxidase may conjugate with more than one IgG fragment, but no fragment can react with more than one peroxidase. The conjugating agent used is N-hydroxy-succinimide ester of N-(4-carboxycyclohexylmethyl) maleimide. This compound is capable of reacting with amino groups *via* one arm and thiol *via* the other arm. The first reaction stage is carrierd in the conventional way by reaction of the amino groups of peroxidase. The result is peroxidase possessing a conjugating reagent capable of reacting with sulfhydryl groups. Rabbit IgG has in its hinge region one disulfide linkage. When this link is reduced, free SH groups are produced and these can be used for conjugation in the second step. Thus rabbit Fab, which possesses a single reduced SH group in the hinge region, is used for conjuction with the maleimide reacted peroxidase. This is the basis of conjugation with maleimide.

The ELISA is done in laboratories using a micro-ELISA plate or other polystyrene microtitre trays. The antigen is diluted in carbonate-bicarbonate buffer to give a concentration of 5-10 μg ml^{-1} which works well for most antigens. 200 μl of the antigen solution is added to each well of a micro-ELISA plate and incubation done at 4°C for 3h in a humid chamber. The antigen is then washed out and the wells are completely filled with PBS-Tween 20 and incubated at room temperature for 3 minutes. The wells are then emptied by inverting and shaking. The washing step is repeated twice. The test serum in which the antigen concentration is to be measured is diluted appropriately and 200 μl of the diluted test serum is added and incubated for 2h at room temperature in a humid chamber. The washing procedure is repeated twice as before. 200 μl of conjugate diluted in PBS-Tween 20 is added and incubated at 4°C overnight in a humid chamber. The washing procedure is again repeated twice. While the plate is incubated, the substrate is prepared by adding 34 mg O-phenylenediamine and 50 ml hydrogen peroxide to 100 ml of citrate phosphate buffer (pH 5.0). This substrate should be made fresh each time and used immediately. 200 μl substrate is added to each well and left in the dark at room temperature for 30 minutes to develop colour. The reaction is stopped by adding 50 μl of suphuric acid (12.5%) and absorbance is read at 492 nm in a spectrophotometer. A calibration curve is plotted using varying concentrations of the test antigen in place of the test serum. From this calibration curve, concentration of the antigen in the test serum can be read. 2, 2-azinodi-3-ethyl benzthiazoline sulphonic acid (ABTS) may be used as an alternative to O-phenylenediamine. In this case, the reaction is stopped by the addition of 50 μl per well of sodium fluoride solution and absorbance read at 650 nm.

Clinical Applications of ELISA

ELISA has numerous applications in the detection and quantitation for several antigens as well as antibodies, both in the clinical laboratory and in biological research. A few of the applications of this technique are listed in Table 4.2 and considered here:

1. VIRAL HEPATITIS A & B - EIAs[77]

Viral hepatitis is a very serious condition and requires diagnosis at a very early stage. Of the different types of hepatitis infections, A and B type are of great significance and can be diagnosed by the use of several serological markers. These include the HAV in stools, IgM antibodies to HAV, total anti-HAV, HBs Ag, HBc Ag, IgM HBc, total Anti-HBC, Anti-HBc, & Anti-HBs. In hepatitis B, the valuable markes are HBs Ag, Anti-HBs, total anti-HBc and IgM HBc, while in hepatitis A, the valuable marker is IgMHAV. Typical serological profiles in the various stages of hepatitis A, hepatitis B (acute and chronic) are as shown in Fig. 4.3. In acute hepatitis, the serological markers initially are HBsAg, IgMHBc and IgMHAV. If these markers are negative, anti-HCV antibodies are looked for. In chronic hepatitis B, HBs Ag, Ig MHBc and total anti-HBc are looked for. For evaluation of the effectiveness of antiviral therapy (interferon therapy), the markers looked for are HBc Ag and anti-Hbc. Also, detection of concomitant HBV and HDV infections by anti-delta antibody is done. Before going in for vaccination, HBs Ag and total anti-HBc, anti-Hbs are measured. If these markers are negative, vaccination can be done. If only HBs Ag is positive, it indicates acute hepatitis and vaccination should not be done in this stage. If anti-HBs and anti-HBs (total) are positive, it indicates that the subject has immunity to HBV and vaccination is not done. After vaccination is done if required, anti-HBs is the serological marker that is measured. This is done to determine the effectiveness of vaccination 1 to 3 months after the injection. A minimum titre of 10m IU/ml should be present in the serum. Serological tests for detection of HBs-Ag, total anti-HBc and AntiHCV are done before blood

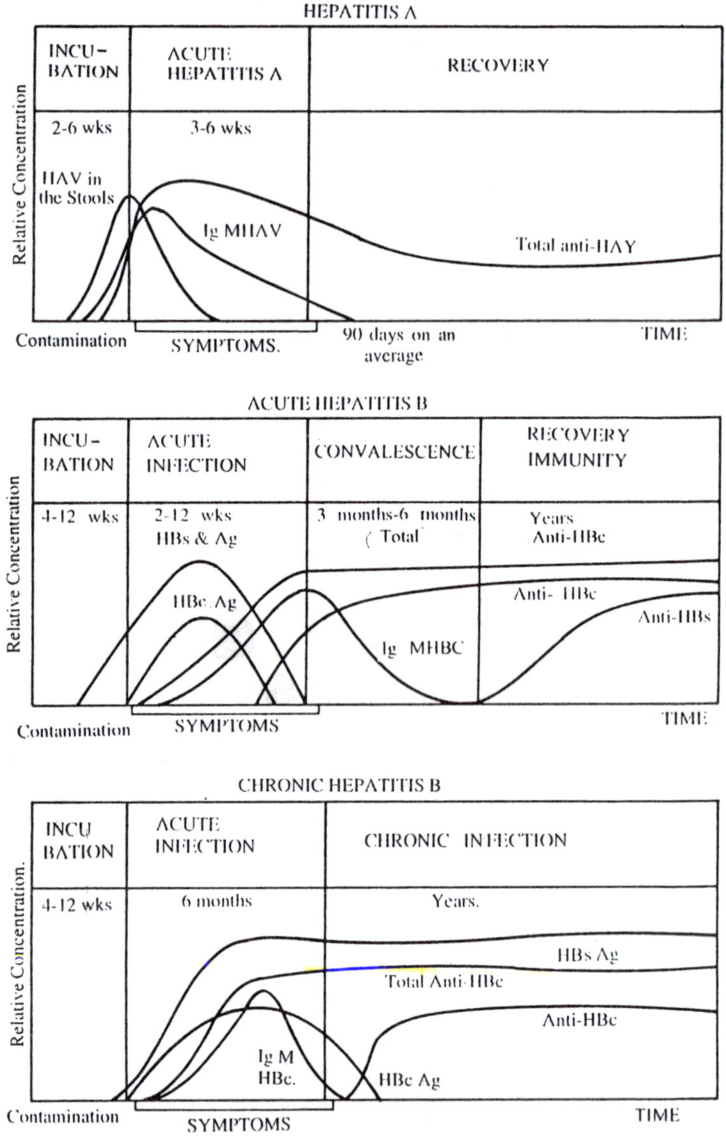

Fig. 4.3. Typical serological profiles in hepatitis A, acute & chronic hepatitis B

transfusion and if one or several of these markers are positive, the blood is not used for transfusion. Screening high risk groups is done by measuring HBs Ag, total anti-HBc, Anti-HBs and Anti-HCV. If all these markers are negative, vaccination is suggested. In case neonate is born to a

HBsAg-positive mother, repeated test for HBsAg and anti-HBs is done during one year and serotherapy and vaccination of the neonate can be done. The interpretation of these serologic tests is given in Table 4.3.

Table 4.2 Clinical applications of ELISA

Condition diagnosed	Parameter detected or quantitated	Type of Assay used
Viral Hepatitis (A and B)	IgM HAV Ab	Immunocapture test
	Total HBc Ab	One-step competition-test
	IgM HBc Ab	Immunocapture test
	HBs Ag	One-step "sandwich" test
	HBs Ab	Two step "sandwich" test
HIV infection	Anti-HIV1	One step "sandwich" test
	Anti-HIV2	One step "sandwich" test
Rotavirus infection	Rotaviral Antigen (Group A)	One step "sandwich" test
Iron deficiency Anemia; Polycythemia; Hemolytic anemia; sickle cell anemia; Thalassemia; etc.	Transferrin receptor	Direct sandwich test.
Chronic gastritis	IgG Abs to *H.pylori*	Sandwich test
Goodpasture's syndrome	Abs to glomerular basement membrane antigen	Sandwich test
Wegener's granulomatosis and related vasculitic diseases.	Anti-neutrophil cytoplasmic Abs	Sandwich test.
Pregnancy	Human Chorionic Gonadotropin (hCG)	Membrane based qualitative ELISA
Colorectal, breast, lung, prostatic pancreatic, & ovarian carcinoma	Carcinoembryonic antigen	Sandwich ELISA (Bead-test)
Hepatocellular carcinoma; Open neural tube defect	Alpha foeto Protein	Sandwich ELISA (Bead test)

Table 4.3 Interpretation of serological tests of hepatitis A & B

IgM HAV	HBs Ag	IgMHBc	Total Anti-HBc	Anti-HBs	Interpretation
+					Hepatitis A
	+				Hepatitis B: early acute phase.
	+	+	+		Hepatitis B: acute infection or less frequent chronic with reactivation.
	+		+		Hepatitis B: chronic carrier (HBs Ag > 6 months)
		+	+		Recent acute hepatitis B: Convalescence phase (Anti HBc Window)
			+	+	Recovery from hepatitis B or, less frequently passive acquisition (transfusion)
				+	Immunization against the HBs Ag (vaccinated subject)
					Look for other infection (HCV, other viral infections).

Clonatec kits are available for performing the assays of these serological markers. Serum IgM antibodies against HAV can be assayed by an immunocapture test (Fig. 4.4a). In this test, an anti-human M chain monoclonal antibody coated on a solid phase is taken and the test serum added. The serum IgM is captured by this antibody. Next, an inactivated suspension of HAV is added which will be captured by the serum IgM. This is followed by the addition of peroxidase-labelled monoclonal antibody against HAV. In between the addition of the reagents in subsequent steps, the plates are washed. 96-well test kits are available and the procedure takes three hours. The presence of specific IgG in the patient's serum, antinuclear antibodies or rheumatoid factors do not influence results.

Serum IgM antibodies against HBc can be similarly assayed by an immunocapture test which is identical to that for anti-HAV but uses a recombinant HBc AG instead of the inactivated HAV suspension. (Fig. 4.4b).

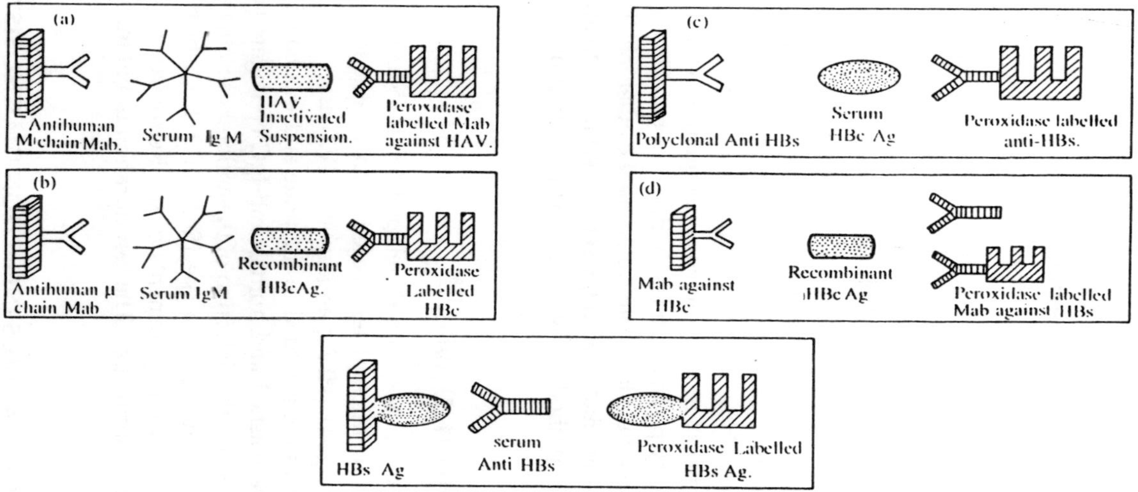

Fig. 4.4. Hepatitis A & B EIA kits.
(a) MHAV Ab EIA: Immunocapture test.
(b) MHBc Ab EIA: Immunocapture test.
(c) HBs Ag EIA: One-step sandwich test.
(d) Total HBc Ab EIA: One-step competition test.
(e) HBs Ab EIA: Two step sandwich test.

HBsAg can be determined in patient's sera by a one step sandwich test (Fig. 4.4c) which is a 96-test kit using a microplate coated with a polyclonal antibody against HBsAg. To the plate is added the serum sample, incubated and washed and then peroxidase labelled monoclonal antibody against HBsAg is added. The test duration is 2 hr 30 minutes. The antibodies used allow detection of all reported types and subtypes of HbsAg.

Total HBc antibodies can be assayed by a one-step competition test (Fig. 4.4 d) which is a 96-test kit using a microplate coated with monoclonal antibody against Hbc. To the plate is added recombinant HbcAg, incubated and washed. Then the test serum and peroxidase labelled monoclonal antibody against HBc are added. The two antibodies compete for the recombinant HBcAg and the measurement of enzyme activity either in the supernatant or the solid phase give a measure of the concentration of serum anti-HBc. The test duration is 1 hr 30 minutes. Anti-HBs levels in serum can be determined either quantitatively or semi-quantitatively by a two-step sandwich test which uses two HBs Ags: one unlabelled and coated on the microplate wells and the other labelled with peroxidase. The plate is incubated with the serum and this is followed by addition of the labelled antibody. (Fig. 4.4e). The test duration is 3 hr 30 minutes.

2. HIV INFECTION -HIV (1+2) AbEIA[78]

An ELISA Sandwich assay has been developed for serodiagnosis and typing of HIV infection. Microplates coated with synthetic peptides homologous to gp 120, gp41 (HIV1) and gp36 (HIV2) are used. A 1:10 dilution of the serum sample is added to the plate and incubated for 60 minutes. The plates are washed and human anti-IgG conjugate labelled with peroxidase is added for the detection of the antibodies in the serum. The kit produced by Clonatec gives a specificity of 99.8% which meets the sensitivity and specificity requirements of the Retrovirus group for the detection of anti-HIV antibodies. The test duration is 120 minutes at laboratory temperature.

3. IDEIA™ KITS FOR ROTAVIRUS INFECTION DIAGNOSIS[79]

The IDEIA™ kit for Rotavirus infection is a simple rapid test procedure for direct detection of Rotavirus in human faeces. It is an EIA utilising a Group A specific monoclonal antibody which is responsible for lack of cross-reactivity with other faecal microorganisms. The IDEIA™ Rotavirus test was tested in collaboration with two independent laboratories on a total of 121 stools from children with gastroenteritis and compared with results obtained by electron microscopy (EM). Discrepant results were resolved by a confirmatory blocking EIA, RNA analysis or serology. The results of the study are given below:

		EM		EM/RNA	
		(+)	(–)	(+)	(–)
IDEIA™	(+)	36	3	38	1
ROTAVIRUS TEST	(–)	0	35	0	35

These results indicate a sensitivity of 100%, specificity of 92% and agreement of 96% of this test with respect to EM and a sensitivity of 100%, specificity of 97% and 99% agreement with respect to the RNA analysis.

4. CLINIGEN™ ELISA FOR HUMAN TRANSFERRIN RECEPTOR[80]

Iron is a trace element recognised as necessary nutrient in man. Iron is required for cell proliferation, by developing reticulocytes in the synthesis of hemoglobin, and by oxidative transport enzymes. The utilisation and storage of iron within the body is complex and highly regulated. Because free iron is toxic to the cell, a system has evolved that utilises soluble proteins to store and carry the iron. Excess iron is stored in tissues as the soluable protein ferritin. Transferrin serves as the transport protein for transfer of bound ferric iron to specific, membrane-bound transferrin receptors (TfR). Research indicates that the expression of TfR correlates to the iron uptake rate and is indicative of the intracellular iron demand. TfR mediates the internalisation of iron into the cell using complex mechanism. The membrance bound Transferrin-TfR is internalised into the cell inside vesicular bodies. The iron is released within these intracellular vesicles and the transferrin-TfR is then transported back to the cell surface within Golgi bodies. The recycled receptor is either incorporated into the cell membrane or shed into the plasma with the apotransferrin.

Circulating levels of TfR reflect the total mass of TfR and their measurement in serum provides a practical means of assessing iron status and effective erythropoiesis. Elevated TfR levels result during erythropoiesis as a function of increased red blood cell production, and from iron deficiency where an increase in the number of TfR per cell results to compensate for the increased intracellular iron demand. Serum TfR levels have been measured in patients with a variety of haematological disorders. Results have indicted elevated levels of TfR in patients with iron deficiency anemia, polycythemia, haemolytic anaemia, sickle cell anaemia, thalassemia and others.[81,82,83,84,85,86,87,88]

R&D systems offers an ELISA test kit for the quantitative determination of transferrin receptor in human serum or plasma. The kit offers six prepared standards that provide a range of 0 to 50 ng/ml. The results are obtained in less than six hours. The sensitivity of the kit is 0.2 ng/ml and shelf life is 9 months. It works on the basic principle of Direct Sandwich ELISA.

5. PYLORAGEN™ -H.PYLORI TEST KIT[89]

Pyloragen™ is a quick, simple serum-based assay to determine the presence of IgG antibodies to *Helicobacter pylori*. *H. pylori* is a gram-negative bacterium that was first identified in the gastric epithelium of patients with active chronic gastritis. Immune responses to *H.pylori* infection have been demonstrated in several laboratories. A number of studies have shown the value of detection of specific IgG for the identification for patients infected with *H.pylori*. There is a correlation between the presence of *H.pylori* and ulcer disease. Also, studies have shown a relationship between the presence of *H.pylori* and gastric carcinoma. Pyloragen™ for IgG antibodies to *H.pylori* uses a microplate ELISA technique. The test has a sensitivity of 97.3%, a specificity of 92.3% and accuracy of 95.6%. A similar kit for IgA antibodies to *H. Pylori* is also available.

6. DIAGNOSIS OF GOODPASTURE'S SYNDROME[90]

Goodpasture's syndrome is an antibody — mediated glomerulonephritis characterised by the presence of antibodies to the glomerular basement membrane (GBM). The resulting immune complexes cause inflammatory reactions, the severity of which determines the degree of renal damage. Common symptoms of this disease are acute or chronic renal failure, proteinuria, oedema, uraemia, anaemia, toxemia, haematuria, haemoptysis and lung haemorrhage. Local and circulating anti-GBM antibodies are specific signs. Mortality in untreated patients is 75-90%. If the disease is detected early enough, renal failure may be avoided or reversed. The treatment of the disease involves suppression of the antibody production. Plasmapheresis with or without immunoadsorption to remove the antibodies is also successful. The key to successful treatment is a fast differential diagnosis of the disease. Conventional diagnosis is based on clinical features, patient history, standard laboratory tests and histopathological examination of kidney biopsies. However, many of the clinical features of Goodpasture's syndrome are also common for other types of glomerulonephritis. Standard laboratory tests add valuable information but are not specific. The presence of kidney lesions (e.g., crescent formation in glomeruli) in conventionally stained histopathological sections gives an indication of the extent of renal damage but is not specific. Traditionally, the examination of kidney biopsies using direct immunofluorescence (IF) has been used to demonstrate the presence of anti-GBM antibodies, specific for Goodpasture's syndrome. However these histopathological examinations involve the trauma of renal biopsy. Indirect method is less sensitive and less specific than direct IF, and is often difficult to interpret. The Biocarb Diagnostics[90] ELISA for detecting anti-GBM antibodies in serum is highly specific because a pure antigen is used. The test is noninvasive and non traumatic for the patient. The method used is a standard ELISA using a maximum of 50 μl of patient serum. The antigen used in the assay is the α3 chain of type IV collagen (M2 subunit) which is specific for GBM. This antigen is used to coat a 48 or a 96 well microtitration plate. For quantitative determinations, three 60 minute incubations are required. The kit includes three standard sera containing antibodies and one negative control. The titre of anti-GBM antibodies in the patient's serum is expressed in ELISA units (EU). A rapid screening kit is also available for which no special equipment is required. Quantitative results are obtained within one hour. Long term use in the department of Nephrology, University Hospital, Lund, Sweden, has shown an excellent correlation between the Biocarb ELISA and direct IF on renal biopsies. All Goodpasture's syndrome patients with linear anti-GBM antibody deposits along the basement membrane had positive ELISA values. The results are shown in Table 4.4. Only 15 out of 559 patients with other renal diseases had borderline positive ELISA values. This test is also useful in following the disease status during treatment. Figure 4.5 shows the antibody titres in two patients undergoing plasmapheresis with Protein A immunoadsorption.

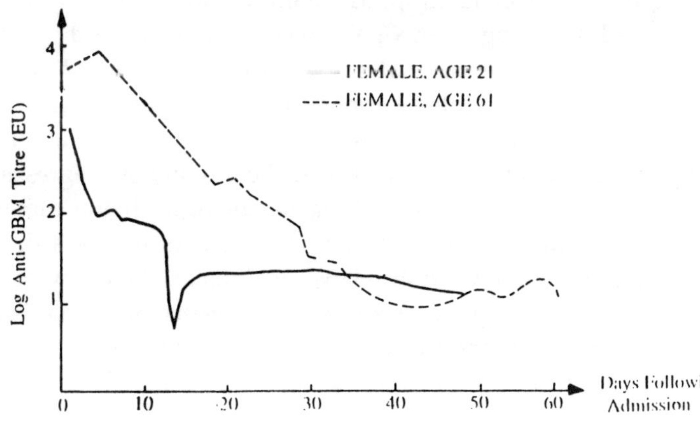

Fig. 4.5. Antibody titres in two patients undergoing plasmapheresis with protein A immunoadsorption

Table 4.4 Performance of Biocarb ELISA

ELISA Titre (EU)	Number of Patients		
	Blood Donors	Goodpasture's	Other renal disease patients
< 10 (Negative)	213	0	544
10-20 (Borderline)	1	3	15
> 20 (Positive)	0	30	0
Total	214	33	559

7. DIAGNOSIS OF WEGENER'S GRANULOMATOSIS AND RELATED VASCULITIC DISEASES[91]

Wegener's granulomatosis (WG) is a systemic necrotising small vessel vasculitis affecting the respiratory tract, ears, kidneys and occasionally other organs. Vasculitis is accompanied by granulomatosis of giant cells in the lungs and leucocyte infiltration of vessel walls. The disease may progress to involve Bowman's capsules resulting in focal necrotising glomerulonephritis. The other related diseases include polyarteritis nodosa (PAN) which is characterised by large vessel vasculitis. When accompanied by small vessel vasculitis, the disease may resemble WG. Microscopic polyarteritis is a non-granulomatous small vessel vasculitis involving the glomeruli. Anti-neutrophil cytoplasmic antibody (ANCA) of C-ANCA specificity is present in approximately 75% of sera from patients with small vessel vasculitis. The serum C-ANCA titre is usually related to the disease activity. These are autoantibodies against the granular components of neutrophils. Immunofluorescence on alcohol-fixed neutrophils using C-ANCA gives a diffuse and granular cytoplasmic staining pattern. C-ANCA appears to be directed against a 29KD protein in the α-granules. Perinuclear staining patterns (P-ANCA) are due to different antigens. Mortality in untreated patients is about 50% within the first six months and is related to the rapidity of progression of the disease and the occurrence of lung hemorrhage. If the disease is detected early enough, renal failure may be avoided or reversed. Since there is a long-term likelyhood of relapse, maintenance therapy or close follow-up for several years is necessary. Treatment with prednisolone and various immunosuppressive drugs has proved to be effective. Plasma exchange and treatement with trimethoprimsulphamethoxazole have been of value in some cases.

The diagnosis of these diseases is based on clinical features, patient history, standard laboratory test and histopathological examination of biopsies. Many of the clinical features of these diseases are also seen in a variety of other glomerulonephritic diseases such as Goodpasture's syndrome, systemic lupus erythematosus and other connective tissue diseases. Standard laboratory tests add valuable information but are not specific. Using conventionally stained histopathological sections, WG giant cells, epithelial cell granulomata and necrotising focal glomerulonephritis. In PAN, there is a large vessel vasculitis arteriography. Biopsies frequently yield non-specific results and thus require repetition delaying diagnosis. ANCA detected by indirect-immunofluorescence (IIF) on alcohol-fixed neutrophils has proved reliable in identifying approximately 75% of WG and related diseases. However, the test requires skilled interpretation. The Biocarb Diagnostics ELISA for detecting C-ANCA correlates well with IIF results and does not depend on subjective interpretation. The method used is a standard ELISA using a maximum of 50 μl of patient serum. The substrate used in the assay is purified material from the α-granules of neutrophils. This material is used to coat a 48 or 96 well microtitration plate. The titre of ANCA is expressed in ELISA units (EU). Long term use in the Department of Nephrology, University Hospital, Lund, Sweden and the Laboratory of Autoimmune Serology, Statens seruminstitut, Copenhagen, Denmark, have shown a good correlation between the Biocarb ELISA and the results obtained from IIF on alcohol-fixed neutrophils. About 75% of patients with active untreated small vessel systemic vasculitis have positive serum ANCA values. Out of1500 patients with renal disorders, 114 patients were C-ANCA positive. All had suspected or verified glomerulonephritis. Table 4.6 shows the advantages of the Biocarb ELISA for C-ANCA over the other techniques used for the diagnosis of Wegener's granulomatosis and related vasculitic diseases. As can be seen from the table, the test is superior to the conventional pathological and biochemical techniques, as well as IIF with respect to ease, speed, specificity and quantifications.

Table 4.5 Comparison of Biocarb ELISA with other diagnostic techniques for Goodpasture's syndrome

Method	Specificty	Speed	Ease	Quantification
Clinical examination	+	+ + +	+ (+)	–
Biopsy-Pathology	+ +	+	+	+
Biopsy-direct IF	+ + +	+	+	+ +
Standard serum lab tests	+ (+)	+ + +	+ + +	+
Indirect IF using serum	+ +	+	+ +	+ +
Biocarb ELISA	+ + +	+ + +	+ + +	+ + +

Table 4.6 Performance of Biocarb ELISA for diagnosis of Wegener's granulomatosis and related vasculitic diseases

Diagnosis	No. of Patients	No. of ANCA—positive patients	
		By IIF	By ELISA
Systemic small vessel vasculitis	72	50	55
Goodpasture's syndrome	15	0	1
Henoch-Schönlein	25	0	1
Systemic lupus erythematosus	40	0	6
Other connective tissue diseases	48	1	3
Total	200	51	66

*IIF: Indirect immunofluorescence

Table 4.7 Comparison of Biocarb ELISA wih other diagnositc techniques for Wegener's granulomatosis.

Method	Specificity	Speed	Ease	Quantification
Clinical examination	+	+++	+(+)	–
Pathology	+ +	+	+	+
Standard serum lab tests	+(+)	+++	+++	+
IIF	+ +	+	+ +	+ +
ELISA	+	+++	+++	+++

8. VISIPREG STRIP PREGNANCY TEST

Human chorionic gonadotropin (hCG) is a glycoprotein hormone produced by the placenta and found in the urine of pregnant women.[92] The hormone increases rapidly during pregnancy and normally reaches levels in excess of 100000 mIU/ml during the first trimester.[93] The molecular structure of hCG is similar to other glycoprotein hormones normally found in the urine, i.e., Luteinizing hormone (LH), Follicle Stimulating Hormone (FSH) and Thyroid Stimulating Hormone (TSH). However, hCG can be differentiated from others by using an antiserum directed against a specific site on the hCG molecule. The VISIPREG STRIP pregnancy test is rapid, sensitive and accurate antibody-based immunoassay for the visual qualitative detection of hCG in the urine as an indicator of pregnancy. It is a two-site sandwich immunoassay in which antiserum anti-hCG is used both in the conjugate form as well as in the native form. It is qualitative test based on the ELISA principle. Strip containing a dried conjugate of anti-hCG (purple in colour) is dipped approximately in 0.5 ml as per the indicated level line (Fig. 4.6). Wicking up by capillary action of the urine sample on the the strip results in reconstitution of the dried conjugate. If hCG is present in the sample, it will react with anti-hCG conjugate to form a complex of "Anti-hCG conjugate-hCG". This complex migrates forward on the membrane pre-dispersed with antiserum anti-hCG. Because the antibody of this immobilised anti-hCG is able to bind/capture the hCG molecule of the migrating complex, a visible purple band is formed along the exact location of the immobilised anti-hCG antiserum denoting a positive result. If there is no hCG present in the sample, the anti-hCG conjugate will pass through the immobilised anti-hCG band and no coloured line will form showing a negative result. Further up the membrane, past the anti-hCG is a control region consisting of a band of immobilised antiserum. This band will bind all free conjugate and form a coloured line regardless of whether hCG is present in the urine or not. Appearance of the control line assures reagent integrity as well as correct testing procedure (Fig. 4.6). For a positive test, 2 lines would thus appear, one, the positive test line and the second one, the control line. However, in a few cases, where concentration of hCG in the sample is very high, only the positive test

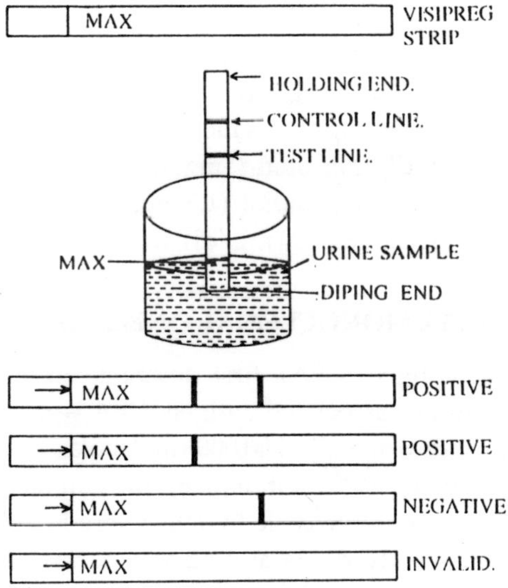

Fig. 4.6. Visipreg strip test for pregnancy diagnosis
(a) A view of the visipreg Strip (b) Performing the test
(c) Interpretation of results.

line will appear and the control line may appear very faint, which may not be visible sometimes (Fig. 4.6). If none of the lines appear, the test is invalid and should be repeated using another strip. Morning urine generally contains the highest concentration of hCG, and urine collected at any time of the day may not be used in the test. The test requires 4 minutes of insulation of the strip in the urine sample. The hCG is not normally detected in the urine of healthy non-pregnant women, In pregnancy, approximately 10 days after implantation, the hCG will reach levels around 50 mIU/ml (using the standard of the World Health Organisation's first International Reference Preparation Irp75/537). The VISIPREG strip will detect 20 mIU/ml of hCG levels. No cross reactivity was detected with the following hormones at their indicated concentrations: LH 500 mIU/ml, FSH 1IU/ml, TSH 1IU/ml. Clinical testing of 915 urine specimens, 640 positive and 275 negative showed 100% correction with a similar qualitative hCG test.[94] Pregnancy is the most probable reason for the presence of hCG in urine, but the possibility of hCG excreting neoplasms[95] should be considered in the diagnosis. Also, the possibility of ectopic pregancy[96,97] must be recognised as a condition in which hCG could also be detected. Sample detected positive by Visipreg strip from very early pregnancy may later prove negative due to natural termination of pregancy. This is estimated to occur in upto 50% of all the conceptions. Hence, it is recommended when using Visipreg Strip to repeat the test after one week. The following substances at the indicated concentration showed no interference in the assay:

Acetaminophen (20 mg/dl), ascorbic acid (20 ml/dl), caffeine, (20 mg/dl), ephedrine (20 mg/dl), gentisic acid (20 mg/dl), phenylpropanolamine (20 mg/dl), salicylic acid (20 mg/dl), EDTA (20 mg/dl), ethanol (1ml/dl), bilirubin (1mg/dl), estriol 17/β (1.4 mg/dl). haemoglobin (1mg/dl), pregnandiol (1.5 mg/dl) atropine (20 mg/ml) cannabinol (20 mg/dl) codeine (20 mg/dl), methadone (20 mg/dl).[94]

9. ABBOTT CEA-EIA MONOCLONAL ONE-STEP

Carcinoembryonic antigen (CEA), first described in 1965 by Gold and Freedman,[98] is a tumor associated antigen. CEA was characterised as a glycoprotein of approximately 200,000 molecular weight with a β-electrophoretic mobility.[99,100] Subsequent development of a RIA by Thomson, *et al.*[101] made it possible to detect very low concentrations of CEA in blood, other body fluids and also in normal and diseased tissues.[102,103,104] Two years later, Hansen *et al.* developed a modified RIA for CEA.[105] The results of clinical studies indicate that CEA, although originally thought to be specific for digestive tract cancers, may also be elevated in other malignancies and in some nonmalignant disorders as well.[106, 107, 108] CEA testing can have significant value in the monitoring of patients with diagnosed malignancies in whom changing concentrations of CEA are observed. A persistent elevation in circulating CEA following treatment is strongly indicative of occult metastatic and/or residual disease.[109, 110]. A persistently rising CEA value may be associated with progressive malignant disease and poor therapeutic response. A declining CEA value is generally indicative of favourable prognosis and good response to treatment. Patients who have low pre-therapy CEA levels may later show elevations in CEA level as an indication of progressive disease. Clinical relevance of the CEA assay has been shown in the follow-up management of patients with colorectal, breast, lung, prostatic, pancreatic, and ovarian carcinoma. Follow-up studies of patients with colorectal, breast and lung carcinoma suggest that the preoperative CEA level has prognostic significance. However, CEA testing is not recommended as a screening procedure to detect cancer in the general population, but use of the CEA as an adjunct test in the prognosis and management of cancer patients has been widely accepted.

"ABBOTT CEA-EIA Monoclonal One-Step" is a solid phase EIA based on the sandwich principle. Samples (specimens, standards, or controls) are incubated simultaneously with anti-CEA (mouse, monoclonal) coated beads and anti-CEA (Mouse, monoclonal) conjugated with horseradish peroxidase (HRP). The solid phase and conjugate monolclonal antibodies recognise different epitopes on the CEA molecule. During this incubation, both immobilised antibody and the enzyme-labelled antibody bind to CEA forming a sandwich. Unbound materials are removed by washing the beads. The beads are then incubated with O-phenylenediamine (OPD) solution containing hydrogen peroxide and a yellow orange colour is produced. The reaction is stopped by the addition of 1N sulphuric acid and the intensity of colour developed is read using a quantum II Analyser set at 492 nm. A standard curve is plotted using the standards in the range 4-80 ng/ml provided in the kit and the concentration of CEA in the sample read from it.

The procedure has some limitations. Elevated circulating CEA levels are common in smokers as well as in patients with nonmalignant disease. Also patients with confirmed carcinoma frequently have a pre-treatment CEA level in the same range as those of healthy/individuals. Specimens from patients who have received preparations of mouse monclonal antibodies for diagnosis or therapy may contain Human anti-mouse antibodies (HAMA). Such specimens may show either falsely elevated or depressed values when tested with assay kits employing mouse antibodies.[111, 112] These specimens should not be assayed with ABBOTT CEA-EIA monoclonal One-step. The distribution of CEA values determined in four institutions (including ABBOTT Laboratories) for 1646 specimens is shown in Table 4.8.[113]

Table 4.8 Distribution of CEA values in four laboratories

		PERCENT %			
	Number	0-3.0 ng/ml	3.1-5.0 ng/ml	5.1-00 ng/ml	> 10.0 ng/ml
HEALTHY SUBJECTS					
Nonsmokers	430	96.5	3.5	0	0
Smokers	184	76.6	18.5	4.9	0
Total	640	90.6	8.0	1.4	0
CARCINOMA					
Colorecteal	180	22.2	3.3	12.8	61.7
Pulmonary	160	41.2	14.4	11.9	32.5
Gastric	28	32.1	21.4	10.7	35.7
Mammary	167	46.7	6.6	13.8	32.9
Pancreatic	33	48.5	15.2	3.0	33.3
Ovarian	81	70.3	11.2	4.9	13.6
Uterine/cervix	7	57.1	14.3	14.3	14.4
NON MALIGANT DISEASE					
Cirrhosis of Liver	42	2.8	26.2	28.6	16.7
Ulcerative Colitis	59	71.2	11.9	10.2	6.8
Diverticulitis	94	83	13.8	3.2	0
Rectal polyps	103	81.6	14.6	2.9	1.0
Pulmonary emphysema	52	50.0	25.0	23.1	1.9

10. ABBOTT AFP-EIA MONOCLONAL

The discovery of Alpha-Fetoprotein (AFP) in fetal serum was first recorded by Bergstrand Czar in 1956.[114] AFP is a single polypeptide chain glycoprotein with a molecular weight of approximately 70,000 daltons. The physico-chemical properties and amino acid composition of AFP are similar to those of albumin. Synthesis of AFP occurs primarily in the liver and the yolk sac of the foetus. It is secreted into the foetal serum, reaching peak at about 13 weeks gestation and gradually declining thereafter. Elevated serum AFP levels subsequently reappear during pregnancy and in conjunction with several malignant diseases. Elevation of serum AFP levels above the normal levels in healthy individuals is seen most notably in nonseminomatous testicular cancer and primary hepatocellular carcinoma. In nonseminomatous testicular carcinoma, a direct relationship has been observed between the incidence of elevated AFP levels and the stage of disease. Elevated AFP levels have also been observed in patients diagnosed as having seminoma with nonseminomatous elements but have not been observed in patients with pure seminoma. hCG and AFP are also important prognostic indicators of survival rate among patients with advanced nonseminomatous germ cell testicular tumors. For patients in clinical remission following treatment, AFP levels generally decrease. Post-operative AFP values, which fail to return to normal, strongly suggest the presence of residual tumor. Tumor recurrence is often accompanied by rise in AFP before progressive disease is clinically evident. More than 70% of patients with primary hepatocellular carcinoma have been reported to have elevated levels of serum AFP. Elevated AFP have occasionally been found in association with gastrointestinal tract cancers with and without liver metastasis and only rarely in other malignancies. Serum AFP has been found to be elevated during pregnancy, in diseases such as ataxia telargiectasia, hereditary tyrosinemia, teratocarcinoma and in benign hepatic conditions such as acute viral hepatitis, chronic active hepatitis and cirrhosis. Many studies have confirmed the utility of AFP in the half detection of open neural tube defect (NTD). In the US, NTD, primarily enecephaly and spina bifida, occur at the rate of between one and two per 1000 live briths and are among the most common major congenital malformations. Encephaly is incompatible with life and accounts for one third to one half of all NTD. Open spina bifida can vary widely in severity. Amniotic fluid (AFP) levels peak at about 13 weeks gestation after which it rapidly declines until about 22 weeks gestation, and then gradually declines until completion of term. Transfer of AFP into maternal circulation is accomplished primarily through diffusion across the placenta. During pregnancy, maternal serum AFP (MSAFP) levels rise from the normal nonpregnant levels of 0-20 ng/ml before 6 weeks gestation to about 250 ng/ml at 32 weeks gestation. If the faetus has an open neural tube defect, AFP is thought to leak directly into the AF causing unexpectedly high values of AFAFP. Subsequently, the AFAFP reaches the maternal circulation, producing abnormally elevated levels of MSAFP. AFP testing during pregnancy is recommended as an effective way to determine those women potentially at risk of carying a foetus with an open NTD. ABBOTTAFP-EIA Monoclonal is a solid phase, two site enzyme immunoassay. The principle and procedure of the assay are similar to that of the CEA assay described earlier. The distribution of AFP values determined in 557 serum specimens is shown in Table 4.9.[115]

Table 4.9 Distribution of serum AFP values

TYPE OF SUBJECT	NUMBER OF SUBJECTS	0-15 ng/ml	15.1-20 ng/ml	20.1-100 ng/ml	100.1-500 ng/ml	Greater than 500 ng/ml
NORMALS	196	196	0	0	0	0
CARCINOMA						
Testicular seminoma	46	45	0	1	0	0
Testicular Non seminoma	130	9	2	40	40	39
Primary hepatocellular	25	1	0	4	11	9
Pancreatic	16	15	1	0	0	0
Ovarian	25	25	0	0	0	0
Gastrointestinal	27	26	1	0	0	0
NONMALIGNANT DISEASE						
Cirrhosis	57	53	0	3	1	0
Hepatitis	35	32	0	1	2	0

Fluoroimmunoassay

Labelled immunoassays employing a fluorophore as label are known as Fluoroimmunoassays (FIA). FIAs can again be of two types: homogeneous and substitutional. The homogeneous FIA is one in which the properties of the label are altered by binding to antibody. These require no separation step and may be both rapid and easily automated. Substitutional FIAs on the other hand require a separation step and can be of different types as in case of RIAs and EIAs.

"Polarisation Fluoroimmunoassay" and "Fluorescence Enhancement and Quenching Assay" are two important homogeneous FIAs used in the clinical as well as research laboratory. Polarisation fluoroimmunoassay is based on the fact that when plane polarised light falls on an antigen labelled with a fluorophore, molecules with their long axes parallel to the plane of light are preferentially excited. If they also rotate between excitation and light emission, the resulting fluorescence is nonpolarised. The degree of depolarisation depends on the rate of rotation and this, in turn, is inversely proportional to molecular size. The binding of a low molecular weight labelled antigen to antibody produces a large complex, which unlike the native antigen, is only capable of rotating slowly so that much of the emitted fluorescene is still polarised. In an immunoassay, unlabelled antigen present in the reaction mixture competes with the fluorescent-labelled antigen for antibody binding sites so that the extent of the depolarisation of the emitted fluorescence is proportional to the concentration of antigen originally present in the solution. Fluorescence polarisation immunoassays are usually far less sensitive than RIAs and limited to the measurement of haptens and low molecular weight antigen, although their sensitivity has been improved by purpose-built instruments. Such assays find application in therapeutic drug monitoring in which blood serum levels of the drug are quantitated. They permit measurement of many therapeutic agents such as gentamicin and phenytoin without prior extraction from plasma. In the fluorescence polarisation assay for, phenytoin, non-specific binding of the fluorescein-labelled ligand to serum proteins occurs causing an increase in the polarisation signal.[116] This interference can be ovecome by proteolytic degradation of the proteins in the serum samples and by the use of purified immunoglobulins isolated from phenytoin antiserum.

Fluorescence enhancement and Quenching immunoassays are based on the fact that when a fluorescent labelled antigen binds to an antibody, it may undergo an alteration in its fluorescent properties. This can take the from of either quenching or enhancement of its fluorescence. Serum gentamicin has been determined by a quenching fluoroimmunoassay technique based on the observation that the intensity of fluorescence produced by fluorescein-labelled gentamicin is reduced by binding to gentamicin antibodies.[117] The effect is diminished by the presence, in the medium of unlabelled gentamicin which competes with the label for antibody binding sites. Differences in fluorescence are fairly small and, as serum samples have an intrinsic fluorescence, this 'blank' value must be measured separately and substracted from the total signal in order to obtain results that correlate well with other assay techniques. No separation phase is required and the assay may be performed using a conventional fluorimeter.

Fluorescence energy transfer immunoassays are based on the principle that when suitable light absorbing molecules are brought into close proximity to appropriate fluorescent molecules showing extensive spectral overlap, there is a transfer of energy between them. Energy transfer occurs within the antigen-antibody complex when, fluorescein labelled antigen is used as a donor molecule and rhodamine-labelled antibody as the acceptor. The quenching of fluorescence by the donor molecule consequent upon its binding to antibody can be inhibited by competition from unlabelled antigen, *i.e.*, the substance to be measured. Such immunoassays are suitable for measurement of both low and high molecular weight compounds.

Luminescent Immunoassay

Luminescent Immunoassays (chemicluminescence and bioluminescence) have potentially high sensitivity. These assays are of various types. In one variant, a covalent complex of a chemiluminescent compound (a luminol derivative), a protein and a steroid hapten was used as labelled antigen analogous to the radiolabel in competitive RIA. The chemiluminescent label was bound by hapten specific antibody and displaced by unlabelled steroid using a second antibody. Chemiluminescence was generated by oxidising the label by hydrogen peroxide and copper acetate at high pH and the emitted light measured in Liquid scintillation counter. The assay as described was sensitive to 0.5 ng testosterone.[118] A homogeneous luminescence assay based on the luciferin-luciferase system which utilises ATP was described by Carrico *et al.*[119] The principle of the assay is given below:

$$\text{ATP} + \text{Luciferin} \xrightarrow[\text{Mg}^{2+}]{\text{Luciferase}} \text{Adenyl Luciferin} + \text{PP}_i$$

$$\text{Adenyl Luciferin} \xrightarrow{\text{O}_2} \text{Adenyl oxyluciferin} + \text{Light.}$$

ATP conjugated to the ligand, 2,4-dinitrobenzene remained active as a substrate for luciferase, but when the conuugate was bound to the ligand-specific antibody the intensity of peak light emission was markedly reduced. The addition of N-(2,4 dinitrophenyl)-β alanine could, by competing with the conjugate for ligand-specific antibody, reduce this inhibitory effect and thereby provided a basis of an assay which, though not as sensitive as some RIAs are for most clinical measurements, in the therapeutic drug monitoring. The sensitivity of the assay could probably be increased by measuring total light emission rather than peak light intensity. Besides, it is important to ensure that samples being assayed do not contain any preformed ATP and that steps are taken to inhibit any enzymes present that are capable of degrading the ATP-ligand bond. Chemiluminescence can be used to increase the sensitivity of enzyme immunoassays. In one example, the activity of a peroxidase label was determined by measuring the chemiluminescence generated by oxidation of luminol by H_2O_2. This improved sensitivity of the assay ten-fold and enabled as little as 10 pg of cortisol to be detected in contrast to a conventional spectrophotometric system which had a sensitivity of 100 pg. The assay contained sufficient interfering substances to prevent the chemiluminescence reaction.[120]

CHAPTER 5

Immunocytochemical Techniques

Introduction

Immunocytochemistry is the localisation of antigens in cells and tissues using a specific immunological reaction. The specificity of high - affinity antibodies not only permits the distinction of different antigen molecules, but also the detection of different epitopes within given antigens. Immunocytochemistry utilises the antibody specificity of immunoglobulins to detect specific constituents within tissue sections prepared for microscopy. Common analytic methods used in biochemistry require isolation of tissue constituent prior to examination. In immunocytochemistry, the tissue constituent is examined *in situ.* There are other methods, besides immunocytochemistry, that permit examination of constituents *in situ.* These methods include enzyme histochemistry in which a substrate analog is reacted with enzyme *in situ* to produce a visible reaction product. However, modern immunocytochemistry is, in general, more specific than enzyme histochemistry. Other nonimmunocytochemical methods include autoradiography, in which a labelled substance is administered *in vivo,* and its distribution as well as that of its metabolites, is examined in tissue sections after prolonged exposure on a photographic plate. Alternatively, a labelled ligand may be applied *in vitro* to a receptor-containing tissue section. These methods are restricted to synthetic or at least highly purified, well defined substances. They often complement immunocytochemical data and can be carried out simultaneously with immunocytochemistry on the same tissue section.

Immunocytochemistry requires specific antibodies. High affinity antibodies increase reaction specificity and reliability and decrease background as well as cross-reactions with undesired antigens. Monoclonal antibodies are best suited to immunocytochemistry and this liberates immunocytochemistry from the need of biochemically defining an antigen prior to immunisation. As in immunochemical methods, the detection of an antigen in immunocytochemistry usually requires the use of conjugated antibodies. Based on the nature of the label, several methods are available and include immunofluorescence, immunoferrtin method, immunogold labelling etc.

Immunofluorescence

Fluorescent dyes such as fluorescein and rhodamine can be coupled to antibodies without destroying their specificity. Such conjugates would combine with antigen present in a tissue section and the bound antibody could be visualised by fluorescence microscopy. Such a test can be used in two ways:

(a) The distribution of antigen throughout a tissue and within cells can be demonstrated.

(b) For the detection of antibodies directed against antigens already known to be present in a given tissue section or cell preparation.

There are three general ways in which the test is carried out.

1. Direct Test

The antibody to the tissue substrate is itself conjugated with fluorochrome and applied directly. By using antisera conjugated to dyes which emit fluorescence at different wavelengths, two different antigens can be identified simultaneously in the same preparation. For example, a mixture of rhodamine labelled anti-IgG and fluorescein labelled anti-IgM when used to stain plasma cells demonstrates that these two classes of antibodies are produced by different cells, One of the medical applications of this technique is the diagnosis of pernicious anemia in which gastric autoantibodies are present in the patient's serum. If the IgG isolated from the patient's serum is fluorescein labelled and used to stain a section of human gastric mucosa on a slide, and viewed under a fluorescence microscope, we would find the cytoplasm of the parietal cells brightly stained. This shows the tissue distribution of the gastric autoantigen. The principle of the direct test is shown in Fig. 5.1a.

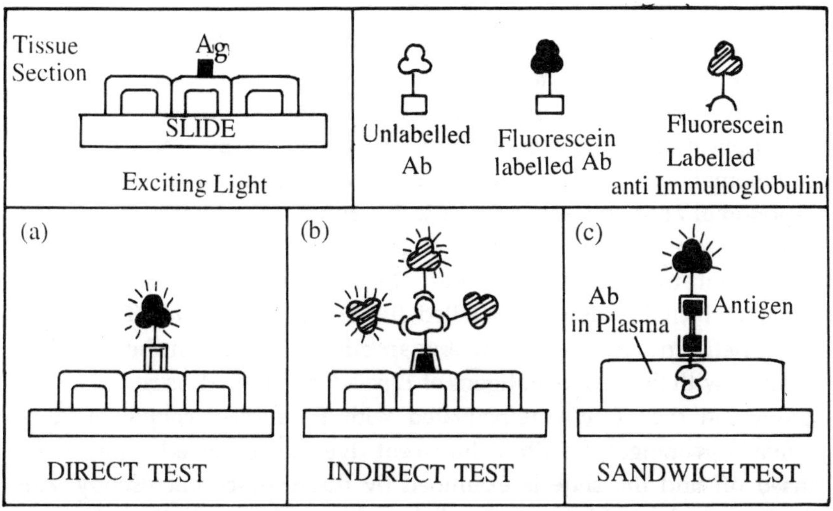

Fig. 5.1. Principle of: (a) Direct, (b) Indirect and (c) Sandwich tests in immunofluorescence

2. Indirect Test

In this technique, the unlabelled antibody is applied directly to the tissue substrate and visualised by treatment with a fluorochrome conjugated anti-immunoglobulin serum. In using this test, the presence of gastric auto-antibodies is to be detected in the patient's serum, the gastric section is first treated with the serum, washed well, and then treated with fluorescein labelled rabbit anti-human immunoglobulin. If antibodies were present, the parietal cells would get stained. The advantages of this technique over the direct test are:

(a) A brighter fluorescence is obtained since several fluorescent anti-immunoglobulin bind onto each of the antibody molecules present in the first layer;

(b) This procedure for the detection of autoantibodies is time saving since there is no need to perform conjugation of every serum sample. The serum sample can be used directly and a standard rabbit anti-human IgG (fluorescein conjugate) can be used for the second layer;

(c) The method is very versatile and can be used for the classing or even subclassing of the autoantibodies in the serum by using conjugates of antisera to individual immunoglobulin heavy chains.

In this method, a third layer can also be used in which case the three layers are: patient's serum, rabbit anti-human-Ig, goat anti-rabbit Ig conjugated with fluorescein. This yields a greater sensitivity but the specificity of the test may go down and so proper controls should be used in the test.

Main application of the indirect test is in the diagnosis of autoimmune diseases such as thyrotoxicosis, insulin dependent *Diabetes mellitus*, Addison's disease, systemic *Lupus erythematosus*, *scleroderma*, etc. One of the applications of the indirect test is the FTA-ABS test which is a treponomal test used in the diagnosis of syphilis. The fluorescence treponomal antibody—absorption test involves initial absorption of the patient's serum with an extract of Reiter treponemes to remove non specific anti-treponomal antibodies. The absorbed serum is applied to a slide on which there is a film of killed, fixed cells of *T. pallidum.* The slide is then briefly incubated to allow any specific antibodies in the serum to combine with the surface antigens of *T. pallidum.* The slide is rinsed free of serum and then briefly reincubated with fluorescent conjugate (anti-human Ig conjugated with a fluorscent dye); uncombined conjugate is rinsed off and the slide is examined by fluorescence microscopy. Antibodies which combine with *T. pallidum* also combine with the conjugate, their presence being indicated by fluorescent treponemes.

3. Sandwich Test

There is a double layer procedure designed to visualise specific antibody. Here, the antigen is sandwiched between the antibody present in the cell substrate and that added as the second layer. If we want to see how many cells in a preparation of lymphoid tissue were synthesising antibody to *Pneumococcus* polysaccharide, we would first fix the cells with ethanol to prevent washing off the antibody during the test, and then treat with a solution of the polysaccharide antigen, after washing, a fluorescein-labelled antibody to the polysaccharide would then be added to locate those cells which had specifically bound the antigen.

The amount of fluorescent antibody bound to each cell can be quantified by a technique known as cytofluorography. In this, the cells are made to flow in a single stream past a laser beam. A light detector at right angles to the beam can be used to record the strength of the fluorescent signal while another detector gives information about cell size by forward light scatter.[121, 122]

Gold Labelling[123]

About 50 years ago fluorescent labels were conjugated with antibodies for the first microscopic detection of antigens in tissues. In 1970, enzyme labels first provided a permanent coloured stain with higher resolution and greater stability than fluorescent signals. With the advent of gold labels in 1971 sensitive high resolution immunocytochemistry became possible in the electron microscope for the first time.[124] The subsequent development of silver enhancing methods then allowed gold labels to provide a highly specific and sensitive method for visualising antigens in light microsopy[125]. Since their introduction to microscopy, gold labels have become recognised as very important for detection and quantititation of proteins, antigens and nucleic acids with other techniques such as blotting, flow cytometry, hybridisation and fingerprint identification.[126]

Gold labels act as markers for molecules that are otherwise invisible to eye or through other detection systems. A colloidal gold conjugate consists of a suspension of gold particles coated with a selected protein or macromolecule (say, antibody). The gold particles may be manufactured to any chosen size from 1-100 nm. This gold probe detection system when incubated with a specific target, such as in a tissue section, will reveal the target through the visibility of the gold particles themselves. In the electron microscope, the particles are visible without further treatment. In the light microscope, however, the gold particles are made visible through a short and simple silver staining procedure. For detection by eye, gold particles will also reveal immobilised protein on a solid phase such as a blotting membrane through the accumulated red colour of the gold sol. Silver enhancement of this gold precipitate also gives further sensitivity detection.

The conjugation of selected proteins to gold particles depends upon at least three physical phenomena:

(a) Charge attraction of the negative gold particle to positively charged protein;

(b) Binding of the gold to sulphur (dative binding) where this may exist within, and

(c) Hydrophobic absorption of the protein to the gold particle surface.

Gold may be conjugated to a wide variety of molecules including proteins, carbohydrates, polysaccharides, enzymes and nucleic acids. The conditions under which a gold conjugate is manufactured drastically affect its performance and stability.

There are several advantages of choosing gold particles as the marker for microscopic and non-microscopic identification of target molecules:

1. The choice of particle size makes it possible to study antigens macroscopically over a wide range of magnifications and provides the opportunity for multiple labelling of antigens on tissue sections.
2. Labelling with gold is completely unequivocal in microscopy. The label cannot be mistaken for any other tissue structure and is either present or not present.
3. While fluorescent labels and enzyme based colour reactions may fade with time and light exposure, gold particles give a permanent label.
4. Because of their particulate nature, quantitative electron microscopy is possible using gold labels, even for more than one antigen by multiple labelling with different particle size.
5. In the light microscope, the intense brown/black stain produced by silver enhancement of the gold signal gives a greater overall sensitivity than the other immunocytochemical methods and counterstaining of the tissue can be done using usual stains.
6. Gold probes are non-toxic, safe and easy to use compared to other labels, they are relatively stable with a longer shelf life.

Gold conjugates may be used directly or indirectly to label antigens. In the direct method, the gold is conjugated to the primary antibody. This allows a single incubation to be performed and provides the simplest detection system. In the indirect method, a primary unlabelled antibody is applied to the specimen to locate the antigen. This is followed by a gold-labelled secondary antibody that detects the primary antibody. The secondary gold-labelled antibodies used with any chosen primary antibody are given below:

Primary Antibody	Secondary Antibody
Rabbit IgG	Goat anti-rabbit IgG
Mouse IgG	Goat anti-mouse IgG
Mouse IgM	Goat anti-mouse IgM
Mouse IgG+IgM	Goat anti-mouse (IgG+IgM)
Rat IgG	Goat anti-rat IgG
Human IgG	Goat anti-human IgG
Guinea pig IgG	Goat anti-guinea pig IgG
Goat IgG	Rabbit anti-goat IgG
Sheep IgG	Donkey anti-sheep IgG.

The principle of the direct and indirect methods are shown in Fig. 5.2a.

In some applications, background labelling may be a problem due to the attraction of the Fc region of the antibody—gold conjugate to tissue components. Normally, this is blocked by the simple application of normal serum prior to the first antibody. If the problem persists, then gold labelled Fab fragments of antibodies may be used. (Fig 5.2 b, c).

A few of the applications of gold labelling are given below:

1. Gold Conjugates for Light Microscopy

For most purposes, 5 nm gold conjugates are adequate. Where higher labelling intensity is required, or where penetration through cell membranes is necessary (such as for cell cultures or cells in suspension), the choice would be 1 nm conjugates. The 1 nm gold conjugates may be diluted much further for use compared to the 5 nm gold conjugates. In both cases, the silver enhancing kit is used to visualise the signal. Specimens may be prepared for labelling studies in the light microsope in a number of ways:

a) *Post-embedding antigens on tissue sections:* The tissue is fixed and embedded in wax or resin and incubations performed on sections. This method may also be used with cryosections.

b) *Labelling of tissues and cells without embedding and sectioning:* Cells in culture may be labelled without embedding. In this case,

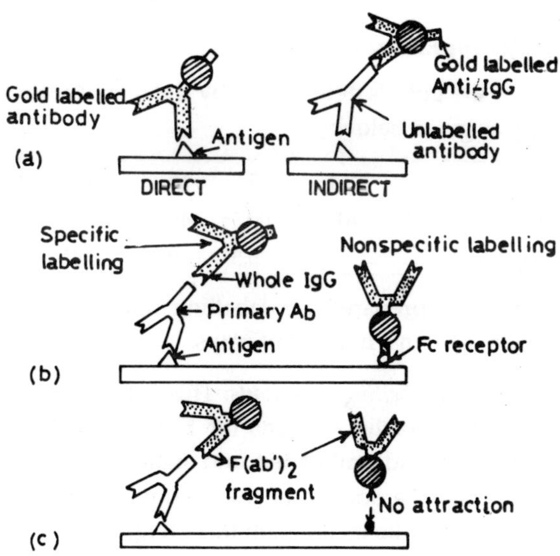

Fig. 5.2. (a) Direct and indirect gold labelling
(b) Demonstration of nonspecific labelling when gold labelled whole IgG is used
(c) Avoiding nonspecific labelling by use of $F(ab')_2$ fragments

penetration of the cell membranes by the antibodies may be necessary for intracellular antigens. The cultures may be observed *in situ.*

c) *Cells in suspension:* Blood cells and cells in suspension may be smeared onto a glass slide and incubated with antibodies. This method is especially useful for examining cell surface antigens.

Specimens for light microscopic examination must usually be mounted on glass slides, and to reduce interaction of the antibodies with the glass itself, slides may be coated with a special tissue adhesive BIOBOND (Biocell Laboratories). This ensures that the specimen remains fixed firmly to the slide throughout. Typical tissue section adhesives such as poly—L-lysine, Elmer's glue, chrome alum, etc., are not suitable for use with immunogold labelling.

Some researchers have occasionally experienced apparent loss of signal one time following immunogold/silver labelling of mounted light microscopic sections. This is due to the formation of translucent silver carboxylate salts from the oxidation of the mounting medium. The silver signal is still there and can be revived by removal of the mountant and reduction of the silver again with photographic developer, but this is very tedious. The British Biocell International offers a new mounting medium, MIOMOUNT, that has low oxidative properties to prevent this problem. It is suitable for both resin and wax sections.

2. Gold Conjugates for Electron Microscopy

The particle size of the gold is determined by the magnification obtained in the microscopy as given below:

Gold particle (nm)	1	5	10	15	20	30
Magnification (screen) x 1000	400	100	40	30	20	10
Particle size (screen) (mm)	0.4	0.5	0.4	0.45	0.4	0.3

Specimens may be prepared for labelling and examination in an electronmicroscope (EM) in a number of ways:

a) *Post-embedding labelling of sections*: This is the most widely used technique for labellig antigens for EM examination. Here the tissue is fixed and embedded and incubations performed on sections. Only the surface of the section is labelled. Ultrathin cryosections may be included in this category.

b) *Pre-embedding labelling of tissues and cells:* In this case, tissues (e.g., cell cultures or cells in suspension) are labelled by incubation, usually after fixation and before embedding. (In electron microscopy, specimen preparation involves fixation followed by embedding in resins like epoxy resins and acrylic resins). This method is particularly appropriate for the identification of antigenic sites on membranes of both isolated cells and microorganisms.

c) *Negative Stain method:* This is the most suitable method for identification of antigens on microorganisms drived onto an EM grid.

d) *Replica technique:* Cell surface antigens on cultured cells may be examined with this method. Cells are fixed and labelled before being replicated for study by EM.

e) *Freeze Fracture technique:* Freeze fracture is stuiable for labelling cell membranes and for examination by thin sections or with replicas. With this technique it is possible to label plasma and intracellular membranes on isolated cells or within tissues.

Large specimens may be examined by Scanning Electron Microscopy (SEM). Specimen surfaces are labelled by preembedding method and then prepared for SEM examination. Backscattered electrons provide an image of the gold particles on the specimen surface. Large gold particles (>20 nm) may be imaged in the SEM with good resolution while the specimen surface is viewed with secondary electron emission. Alternatively, for higher labelling intensity, smaller (e.g., 5 nm) gold conjugates are used and subsequently silver enhanced.

3. Gold Conjugates for In situ Hybridisation

Specific DNA or RNA fragments may be readily identified within cells after hybridisation *in situ* with appropriate DNA/RNA probes. During hybridisation at the cellular level, a labelled specific DNA/RNA probe locates onto complementary cellular target DNA/RNA. This probe is then detected with a suitably labelled antibody.

Isotopically labelled probes have been widely used but suffer a number of disadvantages for routine use in microscopy. The use of a nonisotopic probe is very desirable to avoid hazardous materials and to obtain sensitive, fast, high resolution labelling for LM and EM applications. Antibody based detection methods have proved to be as sensitive as isotopic labels. Enzyme-labelled antibodies have become pupular as safe alternatives to isotopic methods. More recently immunogold techniques have become pupular for both light and electron microscopic detection of DNA/RNA in cells because of their sensitivity and high resolution.

Human papilloma virus has been detected in the light microscope using biotinylated probes followed by goat anti-biotin gold conjugates[127,128] or using digoxigenin-labelled probes followed by sheep anti-digoxigenin gold conjugates.[129] Human papilloma virus and cytomegalvirus have also been detected in cytospins using 1 nm gold conjugates.[130] *In situ* hybridisation may be applied to sections embedded in paraffin wax or resin, or to cells smeared or spun onto glass slides. In the electron microscope, resin embedded sections or whole chromosome spreads may be examined for *in situ* hybridisation studies. The great advantages of gold labelling of nucleic acids by the technique are the high resolution and sensitivity that can be achieved.

The different ways in which gold conjugates can be used for *in situ* hydridisation are shown in Fig. 5.3. A typical protocol for *in situ* hybridisation involves the following steps:

1. Specimen preparation
2. Specimen pretreatment/prehybridisation
3. Preparation of labelled DNA/RNA probe
4. Denaturation of target and probe
5. Hybridisation on section
6. Post-hybridisation washes to remove unbound DNA
7. Blocking of sections against non-specific labelling
8. Incubation with gold labelled antibody
9. Washing
10. Silver enhancement (For EM)

The DNA/RNA probes may be labelled with a number of small reporter molecules such as biotin or digoxigenin. Following hybridisation onto the specimen, the labelled DNA is further detected by antibodies to the reporter molecule. This may be with direct or indirect immunogold methods. For electron microscopy, the particle size is determined by the magnification to be used for subsequent observation as described earlier. Usually ultra small (1 nm) gold conjugates are used and they may be visualised after silver enhancement for 3-5 minutes. For light microscopy, 1 nm and 5 nm conjugates are usually used. For detection of low levels of nucleic acid sequences, 1 nm conjugates are recommended while for many applications, 5 nm conjugates are routinely used.

Use of gold conjugates for *in situ* hybridisation

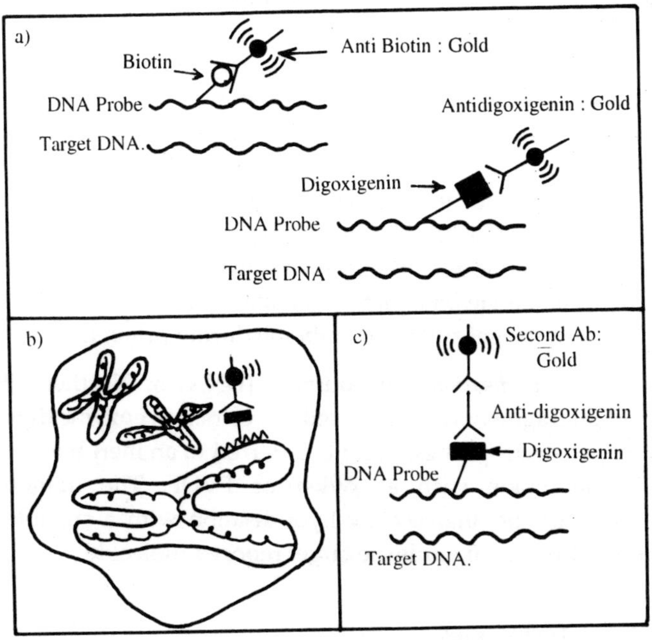

Fig. 5.3. (a) Direct detection of biotin or digoxigenin labelled DNA probes on tissue sections
(b) Direct detection of DNA probe on whole chromosome spread
(c) Indirect detection of DNA probe in tissue section

4. Gold Conjugates for Blotted Proteins and Nucleic Acids

Immunogold labelling is an extremely sensitive method for identifying proteins or nucleic acids which have been immobilised on a membrane such a nitrocellulose or polyvinyl difluoride (PVDF). The technique can be applied to the determination of protein bands transferred from electrophoretic gels, DNA fragments hybridised on membranes with nucleic acid probes, or the identification and quantitation. Blotting techniques have been established for a number of years but the use of immunogold conjugates has provided a safe, sensitive and economical alternative to the traditional labels previously employed.

a) *Proteins*: Blotting applications of gold conjugates include the demonstration of macromolecules, antigens, antibodies, and other proteins immobilised by a suitably negatively charged membrane such as nitrocellulose. When previously separated from a protein or tissue homogenate by electrophoresis and blotted from gels (Western blotting), the individual protein bands may be identified by incubation of the membrane with antibodies directed towards the specific antigen and followed by gold labelled second antibodies. The extremely low background produced by this technique, with high dilutions of the primary and secondary antibodies, ensures crisp definition of the bands and accurate identification. The specific labelling produced in this way can then be compared with a total protein stain obtained from a duplicate blot with PROTOGOLD reagent (a stabilised gold sol specially developed by Biocell laboratories for intense staining of total proteins immobilised on to membranes via Western blotting or direct dot blots). This is a much more accurate technique for protein identification on membranes compared with the coomassie blue method for staining gels where distortion can occur between the gel and the membrane during prtein transfer.

b) *Nucleic Acid Hybridisation*: For Southern (DNA) or Northern (RNA) blots, individual fragments are produced by restriction enzyme digestion, separated by gel electrophoresis, and transferred to an inert but positively charged membrane such as nylon or PVDF. The immobilised fragments can then be analysed by hybridisation with a suitable gold labelled probe, directed towards the target reporter molecule.

Cell Separation Techniques

Certain useful gross separations of immunological cells can be done by physical methods. For example, adherence to plastic surfaces largely removes phagocytic cells while passage down nylon-wool columns greatly enriches lymphocyte populations for T-cells at the expense of B-cells. Lymphocyte preparations are customarily made by density gradient centrifugation which separates them from other peripheral cells of higher buoyant density.

As already described under Immunofluorescence, cell separation can be done in a cytofluorimeter. This is used to separate cells stained with fluorescent antibodies. After flowing past the laser, the liquid stream is broken up into droplets each containing a single cell and if one wishes to separate the stained cell population, the droplet containing a cell giving a fluorescent signal is charged and deviated from the stream by an electrical field into a suitable collector. (Fig. 5.4). This is known as a Fluorescence Activated cell sorter (FACS).

Cells bearing antibody can also be selected by 'Panning' to dishes coated with anti-immunoglobulin. Another technique involves the formation of rosettes with anti-immunoglobulin coated erythrocytes, which can

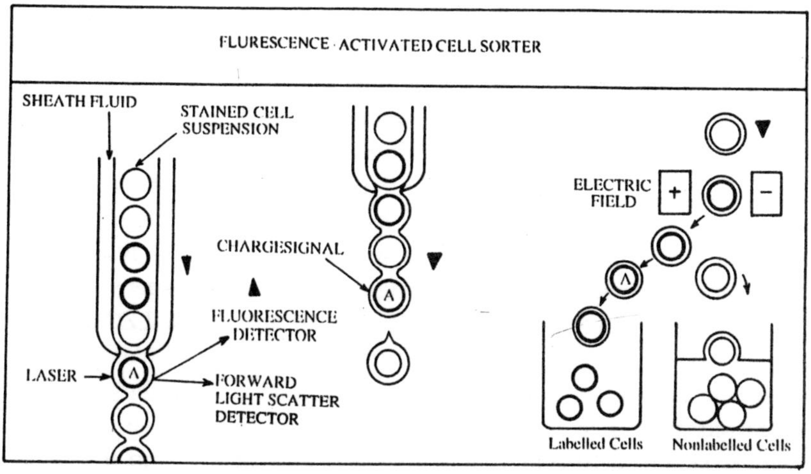

Fig. 5.4. Principle of FACS for flow cytofluorographic measurement of fluorescence on stained cells (rimmed circles) from unstained cells (unrimmed circles). The charge signal can be activated to separate cells of high from low fluorescence or of large from small size

Source : Essential Immunology, I.M. Roitt; ELBS, 1989

be easily separated from unrosetted cells by centrifugation. This method is used for the fractionation of T4 (helper) from T8 (cytotoxic/suppressor) human lymphocytes or the corresponding L3T4 and LY2 mouse populations.

A recently developed technique for the fractionation of cells bearing specific surface antigens from culture is the immunomagnetic method. It is not only applicable for cell fractionation but also for protein purification from a heterogeneous mixture of several proteins. Dynabeads-M-280 immunomagnetic separation technology offers protein separation with reproducible results and reduced nonspecific binding resulting in fewer interference bands and lower background. Dynabeads-M-280 products are uniform superparamagnetic polystyrene beads that provide a magnetic solid phase. Dynabeads bind to the ligand-target molecule forming a complex that can be readily separated from the heterogeneous sample by exposure to a magnet. Elution of the pure captured protein is performed using standard elution procedures.

Immunosensors

Immunocytochemical tests can be fully automated and performed using an image sensing system that allows quantitative analysis with high degree of accuracy and sensitivity. A specific example for the detection of tumour cells will be described here. The specific detection of Line-10 hepatocarcinoma (L-10) cells is based on the complement-mediated cytotoxic reaction with monoclonal antibody 3C4. Cell lysis is followed by using a phase-contrast microscope. In the case of normal cells, the difference between intracellular and extracellular compartments produces a phase-lag of transmitted light because they cannot be lysed. Therefore, normal cells look bright under the phase contrast microscope, whereas damaged cells appear darker because they have lost membrane. Applying the principle of decrease in brightness of the lysed cells, specific and nonspecific cells are distinguished. In actual experiment, 100 µl of cell suspension containing Line-10 cell and nonspecific LZC cell, 100 µl of 3C4 antibody solution and 50 µl of rabbit serum (complement source) are mixed. After 30 minutes of incubation at 37°C, the sample is transferred into a haemocytometer and analyzed by using an image intensifier system. L-10 cells can be detected quantitatively in the range 10-100%, by virtue of producing a linear relationship between the L-10 content and image area in this range. This process does not require a cell washing step, making it possible to automate the whole system during the reaction stage. But the detection limit of the system is not sufficient for cancer diagnosis. So, target cells are stained directly. By using the DNA (or RNA) specific fluorescent dye, propidium iodide, L-10 cells can be detected quantitatively when its content is above 2%.

Immunosensors find application not only in immunocytochemistry but in a variety of other immunoassays. Their introduction has led to the automation of these assays and has made quantitative analysis easier.

CHAPTER 6

Post Script

Advancement in molecular biology in the early sixties saw the emergence of recombinant DNA (rDNA) technology or genetic engineering, which offered tremendous scope to commercialisation of biology in the fields of medicine and agriculture. The commercial utilisation of recombinant DNA technology to obtain beneficial products through manipulation of the genome of the living cells is known as *biotechnology*. The practical potential of biotechnology was so great that both the giant Corporations and the government started thinking in terms of heavy investments to promote research and development in this area.

Achievements

The biotechnology industry, which was spawned about two decades ago, was based on two revolutionising technologies: i) genetic development of monoclonal antibodies, and ii) recombinant DNA technology for gene manipulation. Both technologies have advanced rapidly, and have given birth to a diversity of biogenetic technologies and products. By 1970s all genetic engineering techniques were perfected, and the discovery of restriction endonucleases and DNA ligases made the task easier to construct DNA hybrids. It soon became a reality that through gene splicing experiments, bacteria could be used as factories to produce mammalian proteins. Similarly, it has become possible to produce transgenic organisms with increased resistance to disease, tolerance to environmental conditions, and higher yield.

By 1977, achievements in the use of genetic tricks on the living cells resulted in the expression of commercially important substances, such as interferon, somatostatin, insulin, antibiotics, vaccines, enzymes and immunodiagnostic molecules. It became apparent that the well being and future of mankind lies in practising biotechnology.

With these achievements in hand, new goals were set which included:

1. Protein design — the modification of natural genes to produce novel proteins, capable of acting as powerful pharmaceuticals,

2. Assembly of metabolic pathways for the synthesis of rare and novel chemicals,

3. Production of newer organisms and transgenics,

4. Devising new technologies to produce genetically modified cells to replace wornout and discarded cells,

5. Production of vaccines for AIDS, malaria, hepatitis and other major diseases,

6. Engineering microbes for cleaning up oil spills and toxic wastes, or for producing chemicals from renewable sources.

Application of immunoassays in medical diagnosis began with the measurement of insulin in the plasma using anti-insulin antibodies. The use of monoclonal antibodies held a promise to identify many microorganisms and diseases caused by them. The surge of interest in rapid and accurate diagnosis led to the formulation of enzyme-linked diagnostic kits and reagents used in routine clinical analysis. Other fields exploring enzyme-linked immunoassays are veterinary science, food industry, early pregnancy detection and drug abuse. Hybridoma technology using monoclonal antibodies has opened up new hopes in diagnosis, treatment and monitoring variety of infections and sexually-transmitted diseases.

Potential Hazards from Biotechnology

The benefits that accrue are many, but equally numerous are the hazards that it will create for the mankind, if the experiments go haywire. The dangers that loomed large are the everincreasing production, in research laboratories of mutant viruses, hybrid DNAs with unknown biological potential and adaptation of viruses of human cells. The negative aspect of this technology is, perhaps, the ecological catastrophe it might bring due to disruption of the nature's balance. We may not even be able to assess the nature's destruction in our life time, but it has a vast transformational potential that would be beyond man's control. Genetic manipulation of plants will have extensive genetic erosion and destruction of germplasms.

An alarming possibility that exists is the creation of recombinant DNA molecules that may produce monsters or dangerous materials. Highly toxic chemicals may be produced, which could destroy the whole population of a big city. Accidentally unnatural microorganisms may be produced whose behaviour in the natural environment may wreck havoc and, thus, open up possibilities of biological warfare. It is also possible that a microbe with a specific drug resistance might transform a pathogenic form and create a new disease for the mankind.

Ethical Implications

All these potential hazards create a wide range of consumer, environmental, occupational health and safety issues that directed public attention to the impact of science and technological development. Such anxieties have been fuelled, in particular, by irresponsible behaviour of researchers and medical profession by work related to the human genome and the manipulation of the human embryo which have led to a plethora of moral and legal issues. Bioethics should be regarded as a social movement concerned with ethical values in the formulation of health policy as well as in the practice of biotechnology.

Let us consider the case of immunodiagnostic kits. Quick and reliable kits using monoclonal antibodies were initially developed to ascertain the state of foetus so that mothers could be relieved of their pregnancies on health and medical grounds. These tests have been criticised on moral and legal grounds and the public opinion is both against and in favour. Amniocentesis, a technique employed for the early detection of genetic disorders or abnormalities of the foetus are misused to induce female foeticide. While early detection of pregnancy relieved many a women enabling them to take care of timely health care, it also resulted in foeticide. Legal opinion is divided whether a woman has the right to destroy a healthy foetus.

The case of S.T. and communicable diseases calls for a special mention. Screening for early detection of such diseases in higher risk groups will certainly help checking further spread and giving timely therapy, but such screenings deserve to be kept confidential. HIV infected patients face a frightening and unpredictable future. Social and ethical problems for such patients are formidable. In order to meet this challenge a social code is required. Any test conducted should conform to regulations designed to protect public health and human environment. One should take a future oriented approach concentrated on the need to anticipate ethical dilemmas and develop guidelines at both clinical and research levels.

The genetic engineering issue enhanced the public's role in the direction and management of science and technology, and it became necessary that such techniques should be governed by social and ethical regulations in the interest of human health and society. The hazards of this technology were assessed by an international committee in 1974 under the chairmanship of Lord Eric Ashby, which proposed to impose certain restrictions, and devised a moral code so that multi-faceted risks could be minimised. The committee evolved a code that not only applied to the application of bioengineering but also to the direction of research in this field. Under the garb of strict secrecy and confidentiality, nations did not show inclination to work under controls. To the governments and politicians, rDNA technology became a powerful tool owing to their interest in chemical and biological warfare.

Biotechnology is a double-edged weapon and depends entirely on the manner of exploitation. Science and technology is meant for the good of the society, and drifting from its traditional goals would be disastrous and lead to unknown hazards. There is, therefore, a strong need to bring in protective lesiglation involving the public in science and technology policy making.

Appendix

List of Manufacturers/Suppliers of Immunodiagnostic kits and reagents

Company	Name of the Kit	Tests for
Abbott Laboratories, U.S.A. (Marketing: D.P. Medical Diagnostics Pvt. Ltd., Bombay)	ELISA	–Hepatitis-B –HIV –TSH, FSH, LH, PRL
	RIA	–Tumour antigens
	AFP-EIA-Monoclonal	–Testicular carcinoma
BARC, Bombay	RIA	–FSH –LH
Biocarb Diagnostics Ltd., Sweden	ELISA	–Goodpasture's Syndrome –Wagener's granulomatosis
Biocell Laboratories, U.K.	Gold conjugates	–Determination of Papilloma virus –Proteins and DNA
Biopool, Canada	Minutex	–Fibrin-D-Dimer
Boehringer Knoll Ltd., Bombay	RPHA	–Hepatitis-B
	ELISA	–TSH, FSH, LH
Cadila Laboratories Ltd., Ahmedabad	ELISA	–Filariasis
Clonatec, Paris	ELISA-AbEIA EIA	–HIV –Hepatitis A & B
Dako Diagnostics Ltd., U.K.	IDEA Rotavirus Test	–Detection of Rotavirus infection
Ethnor Ltd., Bombay (Ortho Diagnostic System, Bombay)	ELISA	–Hepatitis-B –HIV
	Latex Agglutination	–Pregnancy –Hepatitis-B
	IHA	–Haeme agglutination
	Blood grouping sera	
Hoechst India, Bombay (Behring Diagnostics Ltd., Bombay)	Latex Agglutination	–Pregnancy
	IHA	
	ELISA	–Hepatitis-B
Hycor Biomedical Inc. U.S.A.	ELISA-Pyloragen	–Detection of IgG antibodies to *H. Pylori*

IHA: Indirect haemeagglutination

Company	Name of the Kit	Tests for
Infar India Ltd., Calcutta	ELISA	–Hepatitis-B –HIV –FSH, LH
J. Mitra & Co. Pvt. Ltd., Delhi	Widal	–*Salmonella* infections
	RPR-Quick Test	–Syphilis
	RF-Agglutinating Quick Test	–Rheumatoid factor
	Visipreg	–Pregnancy
Lupin Diagnostics Ltd., Bhopal	ELISA Ig and Conjugates	–Hepatitis-B –HIV
Merind Ltd., Bombay	ACCUDOT-HIV	–HIV 1 & 2
Miles India Ltd., Baroda	ELISA	–TSH
Pharmacia Ltd., Bangalore	ELISA	–Hepatitis B –HIV –T3/T4
	RIA	–TSH, FSH, LH
Ranbaxy Diagnostics, Delhi	Latex Agglutination	–Pregnancy
	ELISA	–TSH –FSH –Hepatitis-B –HIV
R & D Systems, Minneapolis, U.S.A.	ELISA-Clinigen	–Human Transferrin receptor
Span Diagnostics Ltd., Udahna, Surat	Latex Agglutination	–Pregnancy
	ELISA	–Hepatitis-B –Pregnancy –TSH –T3/T4 –HIV
	RIA Blood grouping sera	
Sri Krishna Laboratories, Ahmedabad	Latex Agglutination	–Pregnancy
	ELISA Latex	–Pregnancy –Hepatitis-B
Wellcome Diagnostics, U.K. (Blue Star Ltd., Bombay)	ELISA	–Pregnancy –Hepatitis-B –HIV
	RIA	

References

1. Roitt, I.M., Molecules which Recognise antigen, *Essential Immunology*, pp. 33-37, ELBS, Oxford (1988).
2. Schiffler *et al.*, *Biochemistry*, **12**, 4620 (1973).
3. Fleet, G.W.J. *et al.*, *Nature*, **224**, 511 (1969).
4. Landsteiner, K. and Vendor Scheer, T., *"Exp. Med."*, **63**, 325 (1936).
5. Hudson, L, and F.C. Hay, "Antibody interaction with antigen". In *Practical Immunology*, pp. 93-141, Blackwell, Oxford (1950).
6. Roitt, I.M., "Antigen Recognition—Primary Interaction" In *Essential Immunology*, pp. 61-63, ELBS, Oxford (1988).
7. Voller, A. and D.E.Bidwell, "Non Isotopic Immunoassays with special reference to Labelled Reagent methods" in *Non Isotopic Immunoassays and their application*, p. 3, Vikas Publishing House, New Delhi (1983).
8. Oudin, T. *C.R. Acad Sci*, Paris, **222**, 115-116 (1946).
9. Glen, W.G., School of Aerospace Med., USAF, Rep No. 62-1, pp. 1-6, (Dec. 1961).
10. Couffer-Kaltenbach, T. and P. Perlman., *J. Biophys. Biochem. Cytol.* **9**, 93-104 (1961).
11. Alfred, J.C., "Single Diffusion Tests" in *Immunodiffusion*, p. 207, Academic Press, New York (1973).
12. Mancini, G. *et al.*, *Immunochemistry*, **2**, 235-254 (1954).
13. Vergani, C. *et al.*, *Immunochemistry*, **4**, 233-237 (1967).
14. Coloren, T. *et al.*, *J. Clin. Pathol.*, **16**, 370-374 (1973).
15. Braum, H.J. and E.W. Aly, *Clin Chim. Acta*, **26**, 588-590 (1969).
16. Fahey, J.L. and E.M. Mc Kelvey., *J. Immunol.* **94**, 84-90 (1965).
17. Vaerman, J.P. and J.F. Heremans, *Immunochemistry*, **6**, 779-786 (1969).
18. Fahey, J.L. and Mc. Kelvey, *Immunochemistry*, **2**, 79-86 (1965).
19. Jones, D., *clin.*, *Chim. Acta*, **29**, 551-556 (1970).
20. Waldenstroms, J., *Diagnosis and Treatment of Multiple Myeloma*, Grune & Stratton, N.Y.
21. Mancini, G. *et al.*, *Immunochemistry*, **2**, 235 (1965).
22. Merler, E., *Immunoglobulins, Biological Aspects and clinical uses*, Washington, D.C., National Academy of Sciences (1970).

23. Ritz Mann, S.E., and J.c. Daniels, *Serum Protein Abnormalities*, Little Browa and Co., Boston Mass. (1975).
24. Hirschfield, T. and U. Soderber, *Experientia* **16**, 198-199 (1960).
25. Croule, A.J., *J. Lab. Clin. Med.*, **55**, 593-604 (1960).
26. Weiner, L.M. *et al., J. Immunol.*, **90**, 788-792 (1963).
27. Schem, R.T. and Rabinovity, M., *Clin. Chim. Acta.*, **13**, 537-538 (1966).
28. Kuwert, E. and B. Niedieck, *Nature*, (London)., **207**, 991-992 (1965).
29. Burstein, M. and Rev. Fr. Etude, *Clin. Biol*, **19**, 918-921 (1969).
30. Catsimpoolas, N., *Immunochemistry*, **6**, 501-502 (1969).
31. Riley, R.F. and M.K. coleman., *J. Lab. clin. Med.* **72**, 714-720 (1968).
32. Micheli, A., *Enzymol. Biol. Clin.* **5**, 175-178 (1965).
33. Daussant, J. and P. Grabar, *Ann. Inst. Pasteur.* **110**, 79-83 (1966).
34. Kaminski, M., *Nature*, (London). **209**, 723-725 (1966).
35. Beerink, E.D. *et al., Immunology*, **9**, 377-389 (1965).
36. Berg, K., *Sangre*, **9**, 24-30 (1964).
37. Slaterus, K.W., *J. Microbiol serol.* **27**, 305-315 (1961).
38. Liu, P.V., *J. Bacteriol.* **81**, 28-35 (1961).
39. Tiu, P.V., Amer. *J. clin. Pathol.* **36**, 471-473 (1961).
40. Mata, L.J. and T.H. Weller., *Proc. Soc. Exp. Biol. Med.* **109**, 705-709 (1962).
41. Gold, P. and S.O. Freedman, *J. Exp. Med.* **122**, 467-481 (1965).
42. Cordoba, F. *et al., Nature* (London) **205**, 565-567 (1965).
43. Van Oss, C.T. and P.M. Bronson, *Immunochemistry*, **6**, 775-778 (1969).
44. *Stained Salmonella Antigens (toxin)* : J. Mitra & Co Pvt. Ltd, New Delhi.
45. *Syphilis RPR Quick test*: J. Mitra & Co. Pvt. Ltd, New Delhi.
46. Portnoy, J. *et al.*, U.S. Public Health Report, **77**, 645 (1962).
47. Stout, G.W. *et al., J. Conf. Pub. Heath Lab* Directors, **26**, 7 (1968).
48. McGrew, B.E. *et al., Amer . J. clin. Path.*, **50**, 52 (1968).
49. McGrew, B.E. *et al., Amer J. clin. Path.*, **50**, 52 (1968).
50. Norins, L.C., *Automation in clinical Chemistry*, **1**, 157 New York Mediad (1968).
51. *Manual of Tests for syphilis*, PHS Publications, No. 411 U.S. Govt. Printing office (1969).
52. Schroeter, A.L. *et al., Adv. in Automated Analysis*, **1**, 256, N. Y. Mediad (1970).
53. Stevens, R.W. and Strrocable, E., J. *clin. path.*, **53**, 32 (1970).
54. *R F — Agglutinating Quick Test*, J. Mitra & Co. Pvt. Ltd., New Delhi.
55. Waaler, E., *Acta. path. Microb. Scand*, **17**, 1 and 2 (1940).
56. Bandilla, K.L. and Mc. Duffie, *Arthritis Rheum.*, **12**, 74 (1969).

57. Muller, W. *The Serology of Rheumatoid Arthritis* Bertin-Gottinger, Heidelberg, p. 97 (1962).

58. Declerk, P.J. *et al., Thromb Haemostas,* **58 (4)** 1024-1029 (1987).

59. Mc Gre, B.E. *et al., Amer. J. clin, Path.* **50**, 52 (1968).

60. Bounameaus, H., *Am. J. Clin. Path.,* **91**:82-85 (1989).

61. Henry P. Treffers, In *McGraw Hill Encyclopaedia of Science and Technology.* Vol. 4. pp. 227-228, McGraw-Hill Inc. USA (1987).

62. Kwaprsti, T.B., *Methodology Immunochemical and Immunologic Research,* (1977), Reprint 1980.

63. Weir, D.M. (ed.), *Handbook of Experimental Immunology,* 3rd ed, 2 vols. (1978).

64. Milgrom, F. and R. Millers., Vox, Sang., **8**, 537-548 (1963).

65. P. Singleton, and Diana Sainstway. *Dictionary of Microbiology and Molecular Biology,* (1987).

66. Sternberger, L.A., *Immunocytochemistry,* pp. 20-24 John Wiley & Sons, N.Y. (1986).

67. Roitt, I.M. *Antigen Recognition-Detection* In *Essential Immunology,* pp. 76-78, ELBS, Oxford (1988).

68. Robinson, J.D. *et al., Radioimmunoassay in Clinical Biochemistry,* Heyden, London (1975).

69. Binoux, M.A. and W.D. Odell, *J. Clin, Endocrinol & Metabolism,* **36**, 303-310 (1973).

70. Kahn, D. *et al., Immunochemistry,* II, 327-332 (1974).

71. Sternberger, L.A. *Immunocytochemistry,* pp-238-239, John Wiley & Sons, N.Y. (1986).

72. Rubenstein, K. *et al.,* BBRC, **47**: 846-85 (1973).

73. Rowlegy, G. *et al., Biol. Chem.,* **250**: 3759-3766 (1973).

74. Burd, J.F. *et al., Clinical Chemistry,* **21**:1766-1768 (1975).

75. Arends, J., *J. Immunol Meth.* **25**: 171, (1979).

76. Wilson, M., Nakane, P., *Immunofluroescence and Related staining Techniques., Amsterdam,* Elsevier, p. 215 (1978).

77. *Hepatitis A and B EIA Kits,* Clonatec, Paris.

78. *HIV (1+2) AbEIA,* Clonatec, Paris.

79. *IDEIA^{TM}-Rotavirus EIA;* DAKO Diagnostics, Ltd., U.K.

80. *CLINIGEN^{TM} ELISA for Human Transferin Receptor,* R & D systems, Monneapolis, U.S.A.

81. Cook, J.D. *et al., Brit. j. Haematol.,* **75**: 603-609 (1990).

82. Cook, J.D., *et al., J. Int. Med.,* **226**: 349-355 (1989).

83. Heubers, *H.A. et al., Blood,* **75** (i): 102-107 (1990).

84. Klenow D., AM. *J. Haematol.,* **34**: 193-198 (1990).
85. Kohgo, Y., *et al., Placenta,* **9**: 523-531 (1988).
86. Kohyol, *et al., Jap. J. Med.* **27** (1): 64-70 (1988).
87. Kohgo, Y., *et al., Blood,* **70** : 1955-1958 (1987).
88. Skikne, B.S., *et al., Blood,* **75** (9): 1870-1876 (1990).
89. *Pyloragen*^{TM} Hycor Biomedical Inc. USA.
90. *Diagnosis of Good pasture's syndrome,* Biocarb Diagnostics, Lund, Sweden.
91. *Diagnosis of Wegener's Granulomatosis and related vasculitic diseases,* Biocarb Diagnostics, Lund, Sweden.
92. Braunstein, G.D. *et al., Am. J. Obstet. Gynacol.,* **126** : 678 (1976).
93. Lendon, E.A., *et al., Fertility and sterility.,* **37**:773 (1982).
94. *Visipreg strip Pregnancy Test,* J. Mitra & Co. Pvt. Ltd. New Delhi.
95. Braunstein, G.D. *et al., Ann. Inter Med.* **78**: 39 (1973).
96. Rasor, J.L. *et al., Obstet, Gynaecol.,* **50**: 533 (1977).
97. Brody, S. *et al., J. Clin. Endocrinol. Metab.,* **22**: 564 (1962).
98. Gold, P. and Freedman, S.O., *J. Exp. Med,* **12** 1 : 439, (1965).
99. Krupey, J. *et al., J. Exp. Med.,* **183** : 387 (1968).
100. Krupey, *et al., Immunochem.,* **9** : 617, (1972).
101. Thomson, D.M.P., *et al., Proc. natl. Acad. Sci.,* USA, **64** : 161, (1969).
102. Zamcheck, N., *Adv. Intern. Med.,* **19** : 413, (1974).
103. Go, V.L.W. *et al., Cancer,* **36** : 2346 (1975).
104. Khoo, S.K., *Int. J. Cancer,* **11** : 681 (1973).
105. Hansen, H.J. *et al., Clin. Res.,* **19** : 143, (1971).
106. Steward, A.M. *et al., Cancer,* **33** : 1246, (1974).
107. Oehr, P, *et al., Tumor Diagnostick,* **1** : 40, (1980).
108. Reynoso, G. *et al., J. Am. Med. Assoc.,* **220** : 361, (1972).
109. Zamcheck, N., Symposium on Clinical Applications of CEA and other antigenic markers assays, Nice, France, October 1977.
110. Martin, E.W. *et al., Am. J. Surgery,* **137** : 167, (1979).
111. Prinus, F.J. *et al., Clin. Chem.,* **34** : 261, (1988).
112. Schroff, R.W. *et al., Cancer Research,* **45** : 879-885, (1985).
113. *ABOTT CEA-EIA Monoclonal One-Step* by ABBOTT Labs, diagnostics Division, North Chicago, USA.
114. Bergstrand, E.G. and Czar B., *Scand J. Clinic & Lab. Invest.,* **8** : 174, (1956).
115. *ABBOTT AFP-EIA Monoclonal,* ABBOTT Labs, Diagnostics Division, North Chicago, USA.
116. McGregor, A.R. *et al., Clinica Chimica Acta,* **83** : 161-167 (1978).
117. Smith, D.S., *FFBS Letters,* **77** : 25-27 (1977).
118. Pratt, J.J. *et al., J. Immunol, Meth.,* **21** : 179-184 (1978).

119. Carrico, R.J. *et al., J. Anal. Biochem.,* **72** : 271-282 (1976).
120. Tsuji A. *et al.,* Enzyme labelled Immunoassay of hormones and drugs, watter de Gryter Berlin, Pal S.B. (ed.), pp 327-339 (1976).
121. Jones, P. *et al., J. Immunol.,* **111** : 1334 (1973).
122. Sternberger, A.L. *Immunocytochemistry,* p 46-47, John Wiley & Sons (1986).
123. British Biocell International Catalogue (1992-93).
124. Faulk Wand Taylor G, *Immunocytchemistry,* **8**, 1081-1083 (1971).
125. Danicher G. and Norgaard, J. *J. Histochem. Cytochem.* **31**, 1994-1398 (1983).
126. Hagat MA (ed). *Colloidal Gold Principles, methods and applications,* Vols 1-3, Academic Press, (1989-91).
127. Jackson, P. *et al., Histochemical,* J. **21**, 425-428 (1989).
128. Jackson, P., *J. Clin. Path.,* **43**, 810-812 (1990).
129. Morey, A.L. *et al., J. Pathol.,* **163** (2) 159 A (1991).
130. Volkers, H.H. Microwave irradiation and DNA *in situ* hypridisation. In *The Microwave cookbook of Pathology.,* eds. ME Boon and LP Kok, Coulomb Press, Leyden, **216** (1988).

RECOMMENDED READING

Benacerraf, B. and E.R. Unanue, *Text Book of Immunology*, Williams and Wilkins Co., Baltimore, 1984.

Capra, J.D. and A.B. Edmundson, The antibody combining site, *Sci. Amer*. **236 (1)** : 50-59, 1977.

Chard, T., An Introduction to Radioimmunoassay and Related Techniques: Laboratory Techniques in Biochemistry and Molecular Biology, Elsevier Science Publishers, Amsterdam, 1986.

Edelman, G.M., Antibody Structure and Molecular Immunology, *Science*, **180** : 830-840, 1973.

Eisen, H., Immunology, 2nd ed., Harper and Row, New York, 1980.

Engvall, L. and F.C. Hay, *Practical Immunology*, Blackwell Scientific Publishers, London, 1976.

Fudenberg, H., D. Sites, J. Caldwell, and J. Wells (eds), *Basic and Clinical Immunology*, Lange Medical Publishers, Los Altos, California, 1980.

Golding, J.W., *Monoclonal Antibodies*, Academic Press, New York, 1984.

Habermehl, K.O. (ed), Rapid Methods and Automation in Microbiology and Immunology, Springer-Verlag, New York, 1985.

Hood, L.E., I.L. Weissman, W.B. Wood, and J.H. Wilson, *Immunology*, 2nd ed., Benjamin/Cummings, Menlo Park, Calif., U.S.A., 1984.

Jerne, N.K., The Immune System, *Sci. Amer*., **229 (1)** : 52 - 60, 1973.

Kurstak, E., *Enzyme Immunodiagnosis*, Academic Press, U.S.A., 1986.

Meyer, R.J. and J. H. Walker, Immunochemical Methods in Cell and Molecular Biology, Academic Press, London, 1987.

Milstein, C., From Antibody Structure to Immunological diversification of Immune response, *Science*, **231**, 1261-1268, 1986.

Ngo, T.T., and H.M. Lenhoff (eds), *Enzyme-Mediated immunoassay*, Plenum Publishing Co., New York, 1985.

Paul, W.E. (ed), *Fundamental Immunology*, Raven Press, U.S.A., 1984.

Steele, R.W., *Immunology for the Practicing Physician*, Appleton and Lange, New York, 1983.

Weir, D.M., C.C. Blackwell, and L.A. Herzenberg, (eds), *Handbook of Experimental Immunology*, 4 volumes, Blackwell Scientific, Oxford, England, 1986.

Glossary

Acquired Immune Deficiency Syndrome: An infectious disease caused by HIV retrovirus, characterised by the loss of normal immune system functions, followed by various infections.

Acquired immunity: The ability of an individual to produce specific antibodies in response to antigens to which body has been previously exposed based on the development of a memory response.

Active immunity: Immunity acquired as a result of the individual's own reactions to pathogenic microorganisms or their antigens, attributable to the presence of antibody in response to an antigenic stimulus.

Adjuvants: Substances that increase the immunological response to vaccine and can be added to vaccines to slow down absorption and increase effectiveness; substances that enhance the action of a drug or antigen.

Agglutination: Clumping or aggregation of cells due to the reaction of surfacebound antigens with homologous antibodies.

Allergen: An antigen that induces an allergic response.

Allergy: Is an immunological hypersensitivity reaction caused by an antigen-antibody reaction marked by a physiological response.

Allotypes: Antigènically different forms of a given type of immunoglobulin occurring in different individuals of the same species.

Anaphylectic hypersensitivity: An exaggerated immune response to foreign protein or other substances, involving the degranulation of mast cells and the release of histamine.

Antibodies: Glycoprotein molecules produced in the body in response to the introduction of an antigen or hapten that can react with that antigen. Also known as immunoglobulins, which are part of the serum fraction of the blood formed in response to antigenic stimulus and which react with antigens with great specificity.

Antibody-mediated immunity: Immunity produced by the activation of the B-lymphocyte population, leading to the production of several classes of antibodies.

Antigen: An agent that initiates antibody formation or induces a state of active immunological hypersensitivity and that can react with immunoglobulins that are formed.

Antigen-antibody complex: The molecular combination that results from the reaction between antigen and complementary antibody molecules.

Antisera: Blood sera that contain antibodies.

Antitoxin: Antibody to a toxin capable of reacting with that poison and neutralising the toxin.

Autoimmunity: Immunity or hypersensitivity to some constituent in one's own body, i.e. immune reaction with self antigens.

B-cell: A differentiated lymphocyte involved in antibody-mediated immunity.

B-lymphocytes: White blood cells that are able to produce specific immunoglobulins; their surfaces carry specific immunoglobulin antigen-binding receptor sites.

Blood serum: The fluid expressed from clotted blood.

Cell-mediated hypersensitivity: Delayed hypersensitivity reaction involving T-lymphocytes and occurring 24-72 hours after exposure to the antigen.

Cell-mediated immunity: Specific acquired immunity involving T cells, primarily responsible for resistance to infectious diseases caused by certain bacteria and viruses that reproduce within host cells.

Classical complement pathway: Series of reactions initiated by the formation of a complex between an antigen and an antibody that lead to the lysis of microbial cells or the enhanced ability of phagocytic blood cells to eliminate such cells.

Clonal selection theory: The theory accounts for antibody formation during foetal development in which a complete set of lymphocytes are developed, each lymphocyte containing the genetic information for initiating an immune response to a single specific antigen for which it has only one type of receptor. B cells that react with self-antigens during this period are destroyed.

Complement: Group of proteins normally present in the plasma and tissue fluids that participates in antigen-antibody reactions, allowing cell lysis to occur.

Complement fixation: The binding of complement to an antigen-antibody complex so that the complement is unavailable for subsequent reactions.

Complement fixation test: It measures the degree of complement fixation for diagnostic purposes.

Complex-mediated hypersensitivity: Type 3 hypersensitivity, a reaction that occurs when excess antigens are produced during a normal inflammatory response and antibody-antigen-complement complexes are deposited in tissues.

Conjugated fluorescent dyes: Fluorescent dyes coupled with antibody molecules used to tag antibodies.

Counter current immunoelectrophoresis: A technique based on the immunological detection of substances that relies on the movement of an antibody and an antigen towards each other in an electric field, resulting in the rapid formation of a detectable antigen-antibody precipitate.

Cross-reactive antibodies: Antibodies that react with heterophile antigens, commonly found in sera of individuals with infectious mononucleosis.

Cytotoxic T cells: Class of T cells (lymphocytes) that are able to kill cells as part of the cell-mediated immune response.

Cytotoxins: Substances that are capable of injuring certain cells without causing cell lysis.

Diffusion: The movement of molecules across a concentration gradient from the area of higher concentration to the area of lower concentration.

Double diffusion test: Immunodiffusion test in which both antigen and antibody diffuse toward each other from separate wells cut into an agar gel.

Electrophoresis: The movement of charged particles suspended in a liquid under the influence of an applied electric field.

Enzyme-linked immunosorbent assay (ELISA): A technique used for detecting and quantifying specific serum antibodies based upon tagging the antigen-antibody complex with a substrate that can be enzymatically converted to readily quantifiable product by a specific enzyme.

Enzymoimmunodiffusion test: A test in which an antigen itself is an enzyme, and antigen-antibody complex exhibits enzyme activity; useful in detecting invisible or nonstainable antigen-antibody aggregates.

Etiology: The study of causation of disease.

Epitopes: Specific antibody binding sites on the surface of antigen, also called antigenic determinants.

Fab (antigen-binding fragment): Either of two identical fragments produced when an immunoglobulin is cleaved by papain; the antigen binding portion of an antibody, including the hypervariable region.

Fc (Crystalisable) fragment: The remainder of the molecule when an immunoglobulin is cleaved and the Fab fragment separated. The crystallisable portion of an immunoglobulin molecule containing the constant region; the end of an immunoglobulin that binds with complement.

Hapten: A substance that elicits antibody formation only when combined with other molecules or particles but that can react with preformed antibodies.

Haemeagglutination: The agglutination or clumping of red blood cells.

Haemeagglutination inhibition (HI): The inhibition of haemeagglutination (antibody-mediated clumping of red blood cells), usually by means of specific immunoglobulins or enzymes, used to determine whether a patient has been exposed to a specific virus.

Hepatitis: Inflammation of liver.

Histocompatibility antigens: Genetically determined isoantigens present on the lipoprotein membranes of nucleated cells of most tissues that cause an immune response when grafted on to a genetically disparate individual and thus determine the compatibility of tissues in transplantation.

Hypervariable region: A region of immunoglobulins that accounts for the specificity of antigen-antibody reactions; genetically specified terminal regions of the Fab fragments.

Hypersensitivity: An exaggerated immunological response upon reexposure to a specific antigen.

Idiotypes: Immunoglobulin molecules with distinct variable regions determining the specificity of antigen-antibody reactions.

IgA: Immunoglobulin A; an antibody that occurs primarily in mucous, semen, and secretions such as saliva, tears, and sweat; an immunoglobulin that plays a major role in protecting mucous membrane surface tissues against microbial infection.

IgD: Immunoglobulin D; an immunoglobulin that normally is present on the suface of some lymphocytes, along with IgM, and appears to play a role in lymphocyte activity.

IgE: Immunoglobulin E; an immunoglobulin that normally is present in blood serum in very low concentrations but that becomes elevated in individuals with allergies; an immunoglobulin that attaches to mast and basophil cells and triggers an allergic response when it reacts with allergens in sensitised individuals.

IgG: Immunoglobulin G; the largest fraction of the body's immunoglobulins and a major antibody that circulates through the body; an immunoglobulin that plays a major role in protecting against systemic microbial organisms.

IgM: Immunoglobulin M; a high molecular weight immunoglobulin occurring as a pentamer that is formed prior to IgG in response to exposure to an antigen; an immunoglobulin that is important in the early response to a microbial infection.

Immunity: The relative unsusceptibility of a person or animal to active infection.

Immunisation: Any procedure in which an antigen is introduced into the body in order to produce a specific immune response.

Immunodeficiency: The lack of an adequate immune response due to inadequate B-or T-cell recognition or response to foregin antigens; a lack of antibody production.

Immunoelectrophoresis: A two-stage procedure used for the analysis of material containing mixtures of distinguishable proteins, e.g., serum, using electrophoretic separation and immunological detection.

Immunoadsorption: A method in which an antigen is usually bound covalently on a matrix to produce an immunoadsorbent.

Immunofluorescence: A technique used to detect a specific antigen or antibody by means of homologous antibodies or antigens that have been conjugated with a fluorescent dye.

Immunogenicity: Ability of a substance to elicit an immune response.

Immunoglobulins (antibodies): A class of proteins found in plasma and other body fluids, including all known antibodies; the antibody fraction of serum; the five classes of antibodies are IgA, IgD, IgE, IgG and IgM.

Immunosuppressant: A drug that depresses the immune response.

Immunosedimentation: A two-step technique for anlaysing antigens in a mixture; the mixture is ultracentrifuged on a sucrose gradient, samples taken from successive levels of the gradient and their antigens are then detected by single diffusion immunoelectrophoresis.

Indirect immunofluorescence: A test used in identifying bacteria such as *Treponema pallidum* by adding dead cells to the patient's serum and adding fluorescent anti-immunoglobulin; if the bacteria stain, the test is positive; the test is indirect because the fluorescent antibody reacts with human IgG after the IgG has reacted with the dead bacterial cells.

Isotope: An element that has the same atomic number as another element but a different atomic weight.

Isotypes: Antibodies differing in heavy chain constant regions associated with different classes and subclasses of immunoglobulins.

Liquid diffusion method: A method used for detecting a substance based upon the diffusion of that substance into a medium to achieve the appropriate concentration for a reaction to occur; a method used in serology for detection of antigens and antibodies based upon its ability it achieve a zone of equivalence in which antigen-antibody reactions can occur.

Lymphocytes: Lymph cells.

Lymphokines: A group of biologically active extracellular proteins formed by activated T lymphocytes involved in cell-mediated immunity; the chemical mediators of cellular immunity.

Macrophages: Mononuclear phagocytes; large, actively phagocytic cells found in spleen; liver, lymph nodes, and blood; important factors in nonspecific immunity.

Major histocompatibility complex (MHC): The genetic region in human beings that controls not only tissue compatibility but also the development and activation of part of immune response.

Monoclonal antibody: An antibody produced from a clone of cells making only that antibody.

Opsonisation: The process by which a cell becomes more susceptible to phagocytosis and lytic digestion when a surface antigen combines with an antibody or other serum component.

Paratope: An important site in the antibody molecule that combines with the antigen.

Passive immunity: Short term immunity brought about by the transfer of preformed antibody from an immune subject to a nonimmune subject.

Precipitin reaction: A serological test in which the interaction of antibodies with soluble antigens is detected by the formation of a precipitate.

Primary immune response: The first immune response to a particular antigen that has a characteristically long lag period and a relatively low titer of antibody production.

Radial immunodiffusion: Radial diffusion is based on the principle that a quantitative relationship exists between the amount of antigen placed in a well where the antibody is already incorporated in the gel, and a ring of precipitation results; generally used in the measurement of serum proteins.

Radioimmunoassay: A highly sensitive serological technique used to assay specific antibodies or antigens, employing a radioactive label to tag the reaction.

Rocket immunoelectrophoresis: A technique in which antiserum to a particular antigen is incorporated into an agarose medium and poured over a glass slide, ensuring nonmigration of the antibody. The pattern of immunoprecipitation resembles a rocket or a spike.

Single immunodiffusion: A technique using agar-gel diffusion in which either antigen or antibody remains fixed and the other reactant is allowed to diffuse freely in the gel.

Single linear immunodiffusion: A test in which one of the two reactants, either antigen or antiserum, remains stationary (the internal reactant) while the other (the external reactant) diffuses into and complexes with it.

Solid phase radioimmunoassay: A test which is used to detect both antigen and antibody using labelled anti-human antibodies.

Syphilis: A chronic, communicable, sexually transmitted disease peculiar to humans, caused by *Treponema pallidum* and characterised by a variety of lesions.

T-cells: Lymphocyte cells that are differentiated in the thymus and are important in cell-mediated immunity, as well as in the modulation of antibody-mediated immunity.

Widal test: Agglutination test for the diagnosis of typhoid fever, using antigens from *Salmonella* species.

Index